Let There Be Light

Characterizing Physical Properties of Colloids, Nanoparticles, Polymers & Proteins Using Light Scattering

Bruce Bradley Weiner, Ph. D.

Bruce B. Weiner
129 Windward Drive
Port Jefferson, New York 11777-2331, U.S.A.
bbradleyw58@gmail.com

Because of the dynamic nature of the Internet, any wrong addresses or links contained in this book may have changed since publication and may no longer be valid.

Cover Design: Paul Roesch, Tupee Studios

Cover Pictures: © Shutterstock—Standard License
https://www.shutterstock.com

ISBN: 0-578-42857-1
ISBN-13: 978-0-578-42857-4

DEDICATION

To my wife's memory. Camille Ofelia Serafica Dumlao Weiner was an extraordinary woman, richly deserving of her long and enchanting name. Warm, gracious, and educated, she never failed to welcome all who crossed her path. Her cheery smile won everyone over. The children and I still greatly miss her.

ACKNOWLEDGMENTS

Two colleagues played an important part in my professional career. Dr. Walther Wilhelm Tscharnuter, co-founder of Brookhaven Instruments, taught me a great deal about how a correlator works and about the importance of making instruments that give reliable results before diving into data interpretation. And, Dr. David Fairhurst, who knows more about practical colloid chemistry than anyone I ever met. He introduced me to the uses and abuses of zeta potential in characterizing small particles and is still a fountain of information.

Special thanks are owed to Brookhaven Instruments Corporation for allowing me to use material I originally wrote for use in their instrument manuals and application notes on their website. I am grateful for their cooperation.

Finally, I owe a debt of gratitude to Mark A. Pfister who guided me through the intricacies of getting this manuscript published and for a review of formatting, though any remaining errors are my responsibility.

PREFACE

What you should know before reading this book: A little chemistry, a little physics, algebra, a very little geometry and trigonometry, and a bit of calculus, though all the important answers are shown in algebraic form. It is written at the first-year graduate school level, though a technician can glean quite a bit from the descriptive parts at the beginning of each chapter and the first few Appendices. Researchers new to these fields but practiced in others can also benefit.

The book is about characterizing the physical properties of submicron particles such as proteins, nanoparticles, and polymers when suspended or dissolved in liquids. Characterization includes determination of size, charge (zeta potential), and molecular mass. Detours into rheology of dilute solutions and suspensions using dynamic light scattering and charge on macroscopic surfaces using phase analysis light scattering are included because these same techniques are used in size and charge characterization of fine particles. Particle characterization is the overarching and unifying theme behind the understanding of the properties of proteins, polymers, and nanoparticles, and the definition of a particle will be explored in the first chapter. You need not, however, know a great deal about proteins, polymers, and nanoparticles: a little perhaps, but not a lot. For example, knowing how to measure the size, charge and mass of a protein does not mean you know what the protein is used for. That is left to the experts using the results from the characterization tools described here.

This book is a composite of introductory concepts suitable for use in interpreting results; an intermediate compendium of useful rules in describing results that instruments produce; and, finally, derivations of some equations used in describing measurements.

Chapter I: Spherical Particles: A Unifying Concept

Chapter II: Static Light Scattering: M_W, R_g, A_2

Chapter III: Dynamic Light Scattering: R_H

Chapter IV: Microrheology: η^*, G', G''

Chapter V: Electrophoretic & Phase Analysis Light Scattering: μ_e, ζ

Chapter VI: Surface Zeta Potential: SZP

Chapter VII: Sedimentation & Centrifugation: d_p, PSD

Scattered throughout each chapter are sample calculations, useful experimental hints, and guidelines for interpretation of data. Special attention is given to limitations in applying results. The appendices elaborate in some detail on specifics in each chapter, and the initial five appendices provide a basic introduction to particles and particle size distributions.

Port Jefferson, New York, May 2019, Bruce B. Weiner

Table of Contents

Chapter IV Microrheology: η^*, G', G''

Chapter V Electrophoretic & Phase Analysis Light Scattering: μ_e, ζ

Chapter VI   Surface Zeta Potential: SZP

Chapter VII   Sedimentation & Centrifugation: d_p, PSD

Chapter I: Spherical Particles: A Unifying Concept

I.1 Introduction

How are proteins, polymers, and nanoparticles alike? They are very small (size measured in nanometers), dominated by surface characteristics, and very prominent in modern science. Even though proteins and polymers are molecules, and nanoparticles maybe not be, they are all made from atoms and molecules, so they all have electrons, some more polarizable than others. Polarizability is at the root of light scattering, which will be explored in Chapters II and III.

Quick, what shape comes to mind first when the words small particle is mentioned? Most people say spheres or globular and conjure up a solid sphere. Yet with a little thought you will agree that a raindrop in air is a liquid particle in a gas (an aerosol). Or an oil droplet in water is a liquid particle in a liquid media (emulsion). Particles need not be homogeneous at the molecular level. A red blood cell is a "soft" particle in saline (0.9% wt/vol NaCl). It is nominally seven microns in diameter, shaped like a donut with the hole filled in. It is a very complex particle, consisting of many distinct types of proteins and other biological chemicals as well as many small inorganic molecules including water all surrounded by a lipid membrane. So, it is not homogeneous on the molecular scale. A random coil polymer dissolved in a nonpolar liquid (polystyrene in toluene) is a particle. It forms a polymer solution, not a suspension. But the random coil isn't a sphere. Unless the polymer is highly branched or cross-linked, it isn't even compact. And then there are self-assembly particles. A micelle is the classic self-assembly particle. A polystyrene or other latex particle suspended in water formed by emulsion polymerization is one example that does come close to a solid, homogeneous sphere, the "model" small particle many people think of first. As a solid suspended in a liquid it is called a sol. See Appendix P1 for other examples of what constitutes a particle.

I.2 Why is the Spherical Shape Important?

There are two reasons: thermodynamics and the convenience of spherical geometry in solving differential equations of physics.

The second law of thermodynamics refers to conditions that produce equilibrium: maximum entropy, minimum free energy including that contributed by surface area. The smaller the surface area for a given volume of material, the lower is the surface free energy. Nature seeks this condition. The minimum surface free energy occurs for a given volume when it is a sphere. The calculus of variations can be used to show that a sphere has the minimum surface area per unit volume.

Creating surface costs energy. This is the reason, barring outside forces, droplets are spheres, planets form spheres (ignore bulging at the equator due to angular momentum while cooling), latex particles, monoclonal antibodies and many other objects are spherical. It is also why little droplets and little bubbles coalesce into big droplets and big bubbles, to minimize surface area per unit volume of material, unless there is another force like electrical charge at the interface that provides repulsion. Without a repulsive force, little particles aggregate and then agglomerate into bigger ones. They do this because there is always an attractive force. Even in the case of electrically neutral particles and molecules, van Der Waals attraction (spontaneous dipole-dipole interactions) pushes particles together. More will be said about this in the sections on zeta potential.

Curiously, and conveniently, lots of basic physics is exactly solvable for spherical shapes. It is why we can describe Stokes' Law of Sedimentation, Stokes' Law of Centrifugation, Stokes-Einstein equation for diffusing particles and more in terms of a sphere's size. It is why we refer to a radius[1] or a diameter so very often when describing size. It is not all pervasive, and there are times when a rod or ellipsoid or random coil is required, or when full image analysis is preferred; however, often an equivalent spherical description is enough.[2]

Now that spheres are firmly attached in your understanding of particle descriptions, let us define spherical properties.

I.3 The Geometry of Spheres and the Equivalent Spherical Diameter (ESD)

Spherical geometry is well known. The surface area of a sphere of radius r and diameter d is given by $S = 4 \cdot \pi \cdot r^2 = \pi \cdot d^2$. The volume of a sphere is given by $V = (4/3) \cdot \pi \cdot r^3 = (\pi/6) \cdot d^3$. Only one size parameter, either r or d, is needed to determine surface area or volume. A cube (hexahedron) shares the same attribute: its surface area $S = 6 \cdot a^2$ and its volume $V = a^3$, where "a" is the length of the cube's side. By contrast, a right-angle cylinder of radius r and

---

[1] A particle analyst means diameter when referring to size. Protein chemists mean radius when referring to size, or, confusingly, molecular weight (kDa). Depending on the field you are in, sizing may mean diameter or radius. Not to worry. For thousands of years we have known that d = 2r. In this book, we will use them interchangeably as the occasion calls for it.

[2] Spheres are so prevalent that the Spherical Chicken Story was once discussed at a scientific workshop. Seems a prominent physics protégé was called home to the family farm when his father died. The farm was famous for producing chickens but was failing. In desperation, the protégé asks his mentor, a renowned physics professor, to come and consult on what to do. The professor spent a week studying every aspect of the farm, its suppliers, the employees, weather patterns, and market conditions. Finally, he announced he had a solution to the problem. He gathered everyone in the bunkhouse and turned to the black board stating "First, we assume the chicken is a sphere."

height h has a surface area given by $S = 2 \cdot \pi \cdot r^2 + 2 \cdot \pi \cdot r \cdot h = (\pi/2) \cdot d^2 + \pi \cdot d \cdot h$. The volume of a cylinder is given by $V = \pi \cdot r^2 \cdot h = (\pi/4) \cdot d^2 \cdot h$. Two parameters, r and h, are needed to describe the surface area and volume of a cylinder. The same is true for ellipsoids of revolution, oblate (flattened) and prolate (elongated) spheroids, and all other regular polyhedrons.

How to characterize non-spherical shapes? The most complete answer is image analysis. But there are drawbacks. First, images, unless they are three dimensional, show a two-dimensional image of a particle (until 3-D imaging in real time is inexpensive). Sometimes the shadows on an image can reveal more detailed shape, but it depends on the lighting. Second, for highly non-spherical particles, completely irregular particles, how many size parameters are required to characterize it completely? Here is a subset of a lengthy list of diameters defined by imaging: Martin's, Feret's, Longest Dimension, Maximum Chord, Perimeter, and Projected Area. Image analysis software for particle sizing lists many more. It is up to the user to determine which ones will correlate with any specific particle property. It is not a straight forward task to do so.

I.4 Geometric ESDs and Their Cousins, ECDs

Often, reducing the complicated imaging information to a geometric equivalent spherical diameter produces useful size distribution information. The concept is simple. By counting pixels of known size along the perimeter of even a complicated two-dimensional image yields the perimeter of the complicated shape. Now set that perimeter equal to $\pi \cdot d_c$, the perimeter of an equivalent circle. This yields the equivalent circular diameter, ECD, given by d_c. Or, count all the pixels of known area inside the complicated, two-dimensional image, correcting for pixels straddling the perimeter. Now set the area of the actual image to $(\pi/4) \cdot d_a^2$, where d_a is a different ECD. If you could determine the volume of an irregular particle, you can define the geometric ESD d_v by setting the measured volume to $(\pi/6) \cdot d_v^3$.

The volume of a prolate ellipsoid of revolution is $(4/3) \cdot \pi \cdot a^2 \cdot b$, where a and b are the minor and major axes. Thus, the equivalent spherical radius ESR $= (a^2 \cdot b)^{1/3}$ and the ESD is $2 \cdot (a^2 \cdot b)^{1/3}$.

I.5 Technique-Defined ESDs

For globular particles and even for irregular particles, where imaging is either not required for shape information or is not necessary for quality control, a technique-defined ESD is possible. Consider a dry powder on a mechanical sieve. The width of the mesh (hole) defines size. It might be the diagonal of a square hole, or the side of a square hole, or the diameter of a round hole. But it is defined by the technique, not a measurement of some geometric particle parameter. True, a long cylindrical particle can fit through a sieve opening if by shaking it pops up vertically before falling through; whereas, lying on its side it is too long to fit through the hole. But that is a problem to be solved by run conditions using a sieve (wet or dry sieving, frequency and amplitude of shaking, duration). The particle size is still determined by the mesh size.

For sedimentation and centrifugation, each governed by a different version of Stokes' Law, the calculated ESD is that of a sphere that settles or centrifuges at the same rate as the actual particle. For dynamic light scattering, DLS, the calculated ESD (or ESR, equivalent spherical radius) is that of a sphere that executes the same translational diffusion as the actual particle. In Fraunhofer Diffraction (sadly when applied to particle sizing called Laser Diffraction) or more generally in Mie Scattering, both forms of static light scattering (intensity as a function of scattering angle), the ESD is that of a sphere that matches most closely the measured diffraction or scattering pattern of the actual particles.

So, each of these techniques should have the calculated diameter with a subscript like those shown in Table I-1.

ESD	Measurement Technique
d_s	Sieving
d_{St}	Sedimentation/Centrifugation
d_{DLS}	Dynamic Light Scattering
d_{FD}	Fraunhofer Diffraction
d_{SLS}	Static Light Scattering

Table I-1: Variety of ESDs obtained from different measuring techniques.

Like the geometrically defined ESDs, these subscripts would be useful in suggesting what technique was used and for non-spherical particles, suggesting what biases may occur in the measurement. The simplest example is elongated particles and sieving. Chances are d_s will underestimate the size of the particles if enough shaking allows the small end of the cylinders to go through the openings or will overestimate the size of the particles if too little shaking allows the cylinders to lie flat and not go through the openings.

In sedimentation and centrifugation, the flow aligns elongated particles and the calculated d_{St} is closer to the diameter of the cylinder than its length. With DLS, FD, and Mie Scattering, generally no such alignment occurs. In fact, the particles are tumbling and rotating, and the results are more likely to represent the longer dimension (though that depends in DLS on whether rotational as well as translational effects are analyzed for).

I.6 The "Size" Axis Label

In a size distribution, whether tabular or graphical, the size axis label is often just plain d for the ESD or r for the ESR. Though it should have a subscript indicating what type of geometric or technique-defined ESD or ESR it is, most often nothing is shown, just d (or r) for size. Remember that when looking at size distribution information. It can help you determine what biases there are in the data. And it will remind you that for non-spherical particles there is no reason to expect any of the ESDs to agree on the same sample. Only when the particles are spheres should the various ESDs agree.

It takes a lot of energy, cohesive energy, to produce small particles from large ones and increase the surface area dramatically. Such energy comes in the form of crushing, grinding, shearing, and sonicating. That is the subject of books on surface chemistry, a subject we will only gloss over here. One can also produce larger particles by growing them from smaller molecules and smaller seed particles. In any case, what are the consequences of producing new surface from the point of view of particle characterization?

I.7 The Importance of Surface Area for Small Particles

Historically, the study of colloids came before polymers, proteins, and certainly nanoparticles. They were defined to be small entities (particles) with the shortest and longest dimensions > 1 nm and < 1,000 nm, respectively. (Though there is no fundamental definition for nanoparticles, they are a subset of colloids, and most often defined with the longest dimension ≤ 100 nm and the shortest dimension > 1 nm.) What is special about 1,000 nm? It is approximately the point at which surface properties begin to play a bigger role in the properties of particles than the bulk properties. See Figure I-1.

You will note that the curve in Figure I-1 is the positive half of a hyperbola, $Y = 6/X$. This is easy to show for a sphere by considering the surface area of a sphere, $S = 4 \cdot \pi \cdot r^2 = \pi \cdot d^2$, and the mass of the sphere with density ϱ. Density is defined as the mass of an object divided by its volume. So, mass is the density times the volume, $M = \varrho \cdot (\pi/6) \cdot d^3$. The surface area per unit mass, also called the Specific Surface Area, $SSA = S/M = 6/(\varrho \cdot d)$. This is the equation of a rectangular hyperbola. And it fits exactly the curve shown in Figure I-1 using a density of 1 g/cm^3.

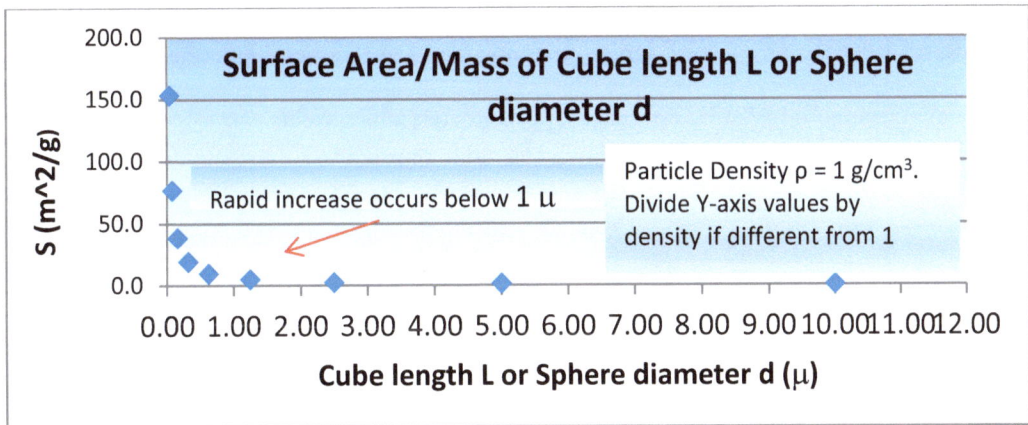

Figure I-1: Plot of specific surface area of a sphere or cube of density 1 g/cm^3 starting with a diameter or side of 10 μ.

The same argument with the same results can be made starting with a cube of 10 μ on a side and slicing it into eight equal cubes with half the original length. See Table I-2 for details for either the sphere or the cube.

X		Y	
Cube Length in μ also Sphere diameter	S in (m)^2/g	No. Cubes or Spheres	No. Cube Sides
10.00	0.6	1.00E+00	6.00E+00
5.00	1.2	8.00E+00	4.80E+01
2.50	2.4	6.40E+01	3.84E+02
1.25	4.8	5.12E+02	3.07E+03
0.63	9.6	4.10E+03	2.46E+04
0.31	19.2	3.28E+04	1.97E+05
0.16	38.4	2.62E+05	1.57E+06
0.08	76.8	2.10E+06	1.26E+07
0.04	153.6	1.68E+07	1.01E+08

Table I-2: Production of X, Y values for use in Figure I-1.

Notice that after eight divisions, from 10 μ to 0.04 μ (rounded), that the number of spheres or cubes has risen from 1 to over 16 million, and that the number of sides has increased to over 100 million. The surface area has grown enormously, going from a single 10 μ object with the same mass as an enormous number of 40 nm (0.04 μ) objects. And the greatest rate of change occurs at or below 1 μ.

Small size means more surface area that needs more wetting, dispersing, or stabilization agent added to counteract attractive forces in order to form dispersions. Gravity is less important, typically, than molecular and electrostatic forces at sizes less than one micron.

Your next idea might be "Why not make a porosimeter, which yields surface area, into a particle sizer and vice versa?" It is not a good idea and here is why:

Why Doesn't a Porosimeter Make a Good Particle Sizer?

What if there were a collection of spheres all with the same density and size? Measure with a porosimeter the surface area per unit mass and calculate the particle size d. Conversely, measure particle size with a particle size analyzer and calculate the surface area per unit mass. But there were two assumptions: spherical shape <u>and no porosity</u>. In fact, a porosimeter is used on dry powders precisely to determine how porous the material really is. And if the measured surface area is much higher than it would be for nonporous spheres, the calculated size d would be much too small. Likewise, given a size d, the calculated surface area is much lower than it would be for porous spheres. Therefore, a porosimeter does not make a good particle sizer and a particle sizer does not make a good porosimeter.

For a collection of assorted size, nonporous spheres, but all the same density, we can calculate what kind of average size is calculated from the SSA. The total surface area S_{tot} is $\Sigma\, n_i \cdot \pi \cdot d_i^2$ and the total mass M_{tot} is $\varrho\Sigma\, n_i \cdot (\pi/6) \cdot d_i^3$. Here d_i is the diameter of the ith size class with n_i particles in it. The sum is over all the classes. The surface area per unit mass is $SSA = 6/\varrho \cdot d_{32}$, where d_{32} is called the surface area average diameter. It is equal to $\Sigma n_i \cdot d_i^3 / \Sigma n_i \cdot d_i^2$. The first subscript, 3, refers to the exponent of d in the numerator and the second subscript, 2, refers to the exponent of d in the denominator.

TIP: Although many particle sizers can calculate a surface area weighted size distribution including d_{32}, if the particles are porous do not pay much attention to the results. And never pay much attention to the value of d_{32} produced by a porosimeter. From the very name itself, a porosimeter is used to characterize total surface area including that of the pores.

I.8 Different Types of Radii Defined

Let R equal the geometric particle radius; R_H equal the hydrodynamic particle radius; and R_g equal the root-mean-square radius of gyration. It is easiest to draw these for a homogeneous spherical particle:

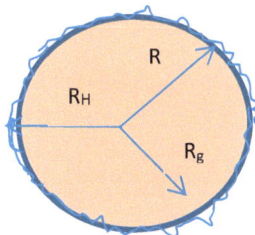

Figure I-2: Geometric radius R, radius of gyration R_g, hydrodynamic radius R_H for a sphere.

Why so many definitions? They each correspond to a different measurement: R from image analysis if there is a sharp edge to define; R_g from Static Light Scattering or Small Angle X-ray and Neutron Scattering; and R_H from Dynamic Light Scattering or Stokes' Law of Sedimentation or Centrifugation.

R_H includes R plus anything attached to and moving with the surface such as adsorbed wetting agents, surfactants, dispersing agents, and long-term stabilizing agents. It may include counterions strongly attracted to a charged surface. In the case of proteins, it includes a layer or so of water molecules. It is everything inside the shear plane, an imaginary layer surrounding the particle wetted by a liquid and about which more will be said in the sections on zeta potential. Often, within experimental error, $R \approx R_H$, except for so-called "hairy" particles stabilized by polymers or long-chain dispersing agents or surfactants, whose length, stretched out from crowding on the surface, is comparable to the particle size. Here is an example:

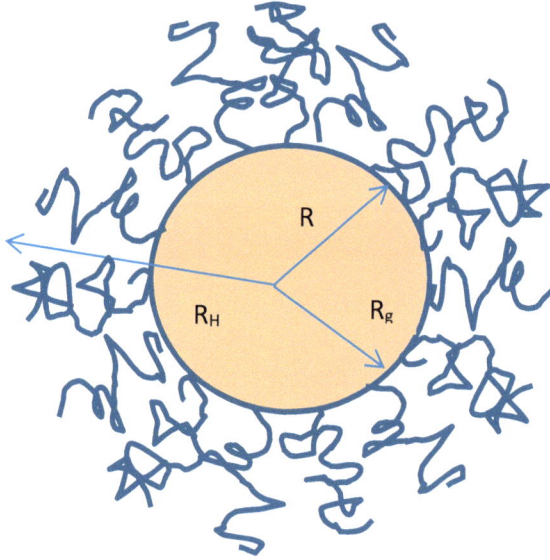

Figure I-3: "Hairy" particle.

In this case, R_H is distinctly larger than R, and R_g depends on the relative densities of core and hairy shell. This case is rarer than that depicted in Figure I-2.

The definition of R_g follows from scattering theory (x-ray, light, and neutron), where little mass elements m_i at a distance r_i from the center of mass of the particle, and the radius of gyration squared is defined as:

$$R_g^2 = \frac{\sum m_i \cdot r_i^2}{\sum m_i} \tag{I-1}$$

Note that the name should have been root-mean-square or RMS radius and not radius of gyration, but the name has stuck for over 60 years and is unlikely to change. More about R_g in Chapter II on static light scattering.

It is important to note the measurement of scattered intensity in static light scattering as a function of scattering angle yields R_g without any assumption of particle shape; whereas, the measurement of the autocorrelation function in dynamic light scattering yields the translational diffusion coefficient. A shape must be assumed and when it is spherical then R_H is calculated from the Stokes-Einstein Law:

$$D_T = \frac{k_B \cdot T}{6 \cdot \pi \cdot \eta \cdot R_H} = \frac{k_B \cdot T}{3 \cdot \pi \cdot \eta \cdot d_H} \qquad (I\text{-}2)$$

Here the subscript T stands for translational but the T in the numerator stands for absolute temperature. The symbol k_B is Boltzmann's constant and η is the liquid viscosity. More about D_T and R_H in Chapter III on dynamic light scattering.

For model shapes of particles and molecules, Table I-3 shows the relationship between R_g and geometric dimensions. (See Reference 4 for details.)

Shape	Sketch	R_g^2
Sphere	→ R	$(3/5)\cdot R^2$
Rod	← → L	$L^2/12$
Random Coil		$b^2 \cdot N/6$

Table I-3: R_g^2 vs geometric size of model particles. N is the number of monomers of length be in polymer chain.

For a solid sphere, $R_g = (3/5)^{1/2} \cdot R = 0.775 \cdot R$ and that is depicted in Figure I-2. Also shown is R_H slightly larger than R because of any counterions, or relatively small stabilizing agents. In most cases $R \approx R_H$.

Figure I-3 depicts a rare case when that is not true. If the sphere is small compared to long chain, stabilizing polymers or large surfactants, the R_H of this "hairy" particle may be significantly larger than R. If both are measurable (R_H from DLS and nanotracking devices, R for electron rich core particles like Au from TEM), then the thickness of the "shell" can be determined. R_g for this case is larger, but not greatly so, because the density of mass elements contributed by the "hairs" is considerably less than that of the core.

An interesting case occurs where the "hairs" are polyelectrolytes. In enough salt, typically > 10 mM for 1:1 electrolytes, the hairs collapse due to the screening of the charges on the polymer backbone. In no salt, they expand due to repulsion of charges along the polymer backbone. The R_H values in each case are very different. Gelatin used to stabilize silver halide particles in the photographic industry provides an interesting example of this phenomenon.

I.9 R_g – R_H and Geometric Dimension Relationships

For spheres, the relationship is $(R_g)^2 = 3/5 \cdot R^2$. R_g is determined from SLS and occasionally SAXS, rarely by SANS. For a spherical polymer[3], biomacromolecule (many globular proteins are close enough), dendrimers, or nanoparticles with small molecules and ions attached to the surface, the geometric radius is replaced by R_H within acceptable limits, provided the model does not conform to Figure I-3. Using DLS, one can determine R_H directly, assuming the correlation function is analyzed correctly for center-of-mass diffusion (translational diffusion). Rods and random coil polymers may add non-translational components to the correlation function and lead to wrong R_H values. (Nanotracking devices yield D_T in two dimensions from which R_H can be determined.)

Thus, for globular shapes, $R_H/R_g = 1/\sqrt{(3/5)} = 1.291$.

For a random coil polymer in a theta solvent at the theta temperature, $R_H/R_g = 0.665$ and in a good solvent, $R_H/R_g = 0.640$.[4] If you must remember these numbers, use 0.65 or 2/3 for both cases. (Theta conditions occur when excluded volume effects are negligible.)

For a long, thin, rod-like macromolecule (prolate ellipsoids with Length/Diameter, L/b, aspect ratios > 10), $(R_g)^2 = L^2/12$. The diameter of the rod plays no part in R_g if the aspect ratio is large enough.

For rod-like macromolecules, including those short enough where the diameter matters:

$$R_H/R_g = \frac{3^{1/2}}{Ln(L/b) - \gamma} \qquad (I\text{-}3)$$

Where, $\gamma = 0.3$ and accounts for end effects of the rod. (See reference 4.) The value of R_H/R_g for a rod is 0.643 when L/b = 20. This value is close to that for random coils. But for smaller L/b, the ratio is larger: 0.865 when the ratio is 10 and 1.323 when the ratio is 5.

---

[3] Polymers are sometimes limited to synthetic polymers. And macromolecules include both polymers and biomacromolecules such as proteins. Throughout this book, the words polymer and macromolecule are often interchangeable.

[4] I. Teraoka, *Polymer Solutions: An Introduction to Physical Properties*, Table 3.1, page 187, John Wiley & Sons, New York (2002).

I.10 R_g – M Relationships

In general, for a given shape, as the molecular weight increases, the size must increase too. How fast it increases depends on the shape of the molecule and the polymer-solvent interactions. In general, $R_g = k\,M^v$, a useful empirical but not absolute relationship. This is one type of Mark-Houwink-Sakurada (MHS) equation first introduced in the 1930's describing the relationship between intrinsic viscosity and molecular weight. These are also called power or scaling laws but apply over limited ranges.

In a so-called "good" solvent, the monomers of a linear, random-coil, homopolymer are surrounded by solvent in a dilute solution. Polymer-solvent contacts are favored over polymer-polymer or solvent-solvent interactions. The extension of the polymer is greatest. In this case, $R_g \sim M^{3/5}$ is a result first given by Flory.[5] (Note to Polymer & Protein Chemists: The second virial coefficient, A_2, is positive under these circumstances.)

An experimental example is polystyrene in toluene or benzene at room temperature from 200 kg/mol to 50 Mg/mol:[6]

$$R_g(nm) = 0.0123 \text{ x (M in g/mol)}^{0.593} \tag{I-4}$$

Such results are obtained by plotting Log R_g vs. Log M_w for either a series of very narrowly distributed polymers, each measurement typically made in batch mode, or using chromatography on a broad sample ($M_w/M_n > 2$) and using each slice to represent a narrow distribution. In either case, it is highly recommended to use at least two full decades in molar mass; otherwise, such Log-Log plots are notorious for yielding the wrong answer. In this case, the range of molar masses was 250:1 or 2.5 decades.

The pre- and exponential factors agree quite well with those given by renormalization theory, 0.01234 and 0.5936 (see reference 4). And the exponent is very close to 3/5 from Flory's simpler theory. Using this formula to estimate R_g for polystyrene in other good solvents yields reasonable approximations. Goods solvents dissolve the polymer readily.

However, the pre-exponential factor and exponent will differ for another linear, random coil, homopolymer. Values in the range of 0.55 to 0.65 are common for the exponent.

In a so-called "theta" solvent, at the theta temperature, various effects (like excluded volume) cancel and the size is proportional to the square root of the number of monomers. That is $R_g \sim M^{1/2}$. The second virial coefficient, A_2, is zero. Thus, the solution behaves as if it were ideal until third order effects (A_3Mc^2) become important. Each polymer/solvent pair

---

[5] P.J. Flory, *Principles of Polymer Chemistry*, Cornell University Press, Ithaca, New York (1953).

[6] J. de Cloizeaux, G. Jannink, *Polymers in Solution: Their Modelling and Structure*, Oxford University Press: Clarendon (1990).

has, in principle, a theta temperature. However, it may be outside the range (in a phase diagram sense) where the polymer/solvent can form a solution.

An experimental example is polystyrene in cyclohexane at 35.4 °C from 800 kg/mol to 50 Mg/mol (see reference 4):

$$R_g(nm) = 0.02675 \text{ x } (M \text{ in g/mol})^{0.5040} \tag{I-5}$$

As the solvent becomes "poor", the polymer-polymer and solvent-solvent interactions become stronger than the polymer-solvent interactions. The second virial coefficient, A_2, is less positive, even somewhat negative. The polymer begins to fold up, driving towards a globular shape. For a sphere, the hydrodynamic volume V_H is $\sim (R_H)^3 \sim (R_g)^3$. The volume is proportional to the mass. Therefore, it follows that $R_g \sim M^{1/3}$ for a sphere.

It is hard to find such cases starting with a synthetic, random coil polymer, for when A_2 is negative and polymer-polymer interactions are strong, they are also strong between chains. Aggregation occurs. Thus, the molecular weight and R_g determined are of the aggregates.

For long, thin rods, $R_g \sim L \sim M$, since adding monomer increases the length and molecular weight linearly. In this idealized case, the exponent for a rod is 1.

The exponent in a Log R_g vs. Log M plot carries shape information. If it is close to 1, the shape is rod-like; if it is close to 0.6 the shape is that of a random coil in a good solvent; if it is 0.5 the shape is that of a random coil under theta conditions; and as the exponent approaches 1/3 the shape approaches that of a spherical particle.

I.11 R_H – M Relationships

It was shown that for a given shape there is a fixed relationship between R_g and R_H. It follows that if DLS is used to determine R_H, then the exponent in R_H-M relationships also indicates something about the shape.

For example, DLS measurements on 14 globular proteins, ranging from lysozyme (14.7 kg/mol) to thyroglobulin (670 kg/mol) at room temperature gave this result:[7]

Globular Proteins: $$R_H \text{ (nm)} = 0.0309 \text{ x } (M \text{ in g/mol})^{0.428} \tag{I-6}$$

Clearly, many of the globular proteins are slightly elongated since the exponent is not 1/3. Given a direct measurement of R_H using DLS, this equation can predict the molecular weight of a globular protein to within 10%. Not bad if the goal is to determine whether the protein has aggregated.

---

[7] K. Mattison, private communication of DLS measurements made with a Brookhaven BI-200SM multiangle SLS/DLS goniometer circa 1995.

How about other biomacromolecules? Table I-4 shows average results in water at room temperature as given in the Brookhaven Instrument's ParticleSolutions software. (Free to download at www.brookhaveninstruments.com. Click on the Protein or Polymer calculators.)

Globular Proteins	R_H (nm) = 0.0309 x (M in g/mol)$^{0.428}$
Starburst Dendrimers	R_H (nm) = 0.0860$_5$ x (M in g/mol)$^{0.30}$
Branched Polysaccharides	R_H (nm) = 0.0367 x (M in g/mol)$^{0.44}$
Linear Polysaccharides	R_H (nm) = 0.0151 x (M in g/mol)$^{0.55}$

Table I-4: R_H vs. M for biomacromolecules.

The exponent for dendrimers confirms that these polymers made from trifunctional monomers are sphere-like. The branched polysaccharides are as dense as a globular protein given the similarity of the exponents. And linear polysaccharides in water behave as many linear homopolymers in a "good" solvent do.

I.12 R_H – D_T Relationships

The relationship between hydrodynamic radius and translational diffusion coefficient for center-of-mass motion is given by the Stokes-Einstein equation:

$$D_T = k_B \cdot T / 6 \cdot \pi \cdot \eta \cdot R_H \qquad \text{(I-7)}$$

Where D_T is the translational diffusion coefficient, T is temperature in Kelvin; k_B is Boltzmann's constant (1.3806 E-23 J·K^{-1}); and η is the bulk viscosity of the liquid (0.8904 mPa·s for water at 298.15 K). The previous four R_H – M relationships were in fact calculated from the four $D_T = k_D \cdot M^\beta$ relationships shown in Table I-5 (all at 25 °C, pure water):

Globular Proteins	D_T(cm^2·s^{-1}) = 7.89E-5 x (M in g/mol)$^{-0.428}$
Starburst Dendrimers	D_T(cm^2·s^{-1}) = 2.86E-5 x (M in g/mol)$^{-0.30}$
Branched Polysaccharides	D_T(cm^2·s^{-1}) = 6.71E-5 x (M in g/mol)$^{-0.44}$
Linear Polysaccharides	D_T(cm^2·s^{-1}) = 1.63E-4 x (M in g/mol)$^{-0.55}$

Table I-5: D_T vs. M for biomacromolecules.

As stated before, a relationship such as $D_T = k_D \cdot M^\beta$ is known as a Mark-Houwink-Sakurada (MHS) equation with the pair of constants, {k_D, β}, known as the MHS constants. A more familiar set of MHS constants is described in the next section. All such MHS equations are empirical. [The SLS equation relating excess scattered intensity to molecular weight is not empirical. It is more fundamental. Thus, the name "absolute" attached to molecular weight determination using SLS, even though the instrument itself must be calibrated.]

This relationship can be used to calculate a molecular weight (not M_n or M_w, but M_D) from DLS measurements of D_T provided the model is adequate to describe the experimental situation.

I.13 [η] – M Relationships

The most famous MHS equation relates intrinsic viscosity, [η], to molecular weight, M, as follows:

$$[\eta] = k_\eta \cdot M^\alpha$$

(1-8)

Where the pair of MHS constants, $\{k_\eta, \alpha\}$, depend on the solvent, the solute (macromolecule), and the temperature. The constants α are related to shape with values >1 for rods, 0.7 to 0.8 for random coils in good solvents, and 0.5 in theta solvents. This relationship can be used to calculate a molecular weight, called the viscosity-average molar mass, M_v, from intrinsic viscosity measurements of [η] provided the model is adequate to describe the experimental conditions.

While this equation was used extensively before size exclusion chromatography/gel permeation chromatography (SEC/GPC) became routine, a more common place to find the use of intrinsic viscosity is this empirical equation that is ultimately derived from the Einstein equation for spheres:

$$[\eta] \cdot M = \Phi \cdot V_H$$

(I-9)

Where V_H is the hydrodynamic volume and Φ is a constant equal to about 2.8 x 10^{23} for all flexible chains.[8] If it were a universal constant, then the product of intrinsic viscosity and molecular weight would be the same for all polymers with the same hydrodynamic volume. Since hydrodynamic volume is responsible for pure size exclusion chromatography, it follows that setting up a curve of [η]·M vs. V_R, where V_R is retention volume in an SEC experiment, should lead to different polymers of different molecular weights falling on the same curve. Then a measurement of the intrinsic viscosity of an unknown would lead to its molecular weight if it too followed the "universal" curve.

Many nonpolar flexible chain polymers and copolymers in nonpolar liquids do indeed, over limited molecular weight ranges, follow such a "universal" curve. And, so, this empirical

---

[8] Paul C. Hiemenz and Timothy P. Lodge, *Polymer Chemistry, Second Edition*, CRC Press, Boca Raton, Florida (2007).

equation provides a method of calibrating columns if standards that match the unknown in structure and chemistry are not available for standard calibration.

But the equation is not universal. Several polymers, especially ones in water and polar or charged macromolecules with retention mechanisms other than size, do not follow the universal curve. The constant Φ is supposed to be independent of structure and shape. But it isn't for many types of polymers. And "universal" calibration is not, after all, quite so universal.

[Note: SEC/GPC when applied to proteins and other biomacromolecules is called gel filtration chromatography, GFC.]

I.14 Summary and Sample Calculations

The spherical model for particles can be used to interpret properties. Experimental measurements of distinct types of radii and diameters not only yields sizes but can, in some circumstances, shed light on shape. A plethora of exponential relationships (MHS equations, scaling laws) can be used to estimate properties, which is useful when looking for consistency.

If the reader is unfamiliar with particle sizing, read Appendices P1 – P5. Discover what is a particle, how size is defined, and how distributions (continuous and discrete) and weighting are defined.

Fun with Spherical and MHS Equations and How to Get into Trouble

1. Given a DLS measurement of 92 nm for a standard polystyrene latex sphere in water with density 1.051 g/cm³, what are R_H, R_g, and the pseudo molecular weight M?

    $d \approx d_H = 92$ nm so $R_H = 46$ nm and $R_g \approx 0.775 \cdot R_H = 35.6$ nm. Test by measuring with SLS over angular range of say 30° to 150° scattering angle.

    $m = \varrho \cdot v = 1.051$ g/cm³$\cdot(\pi/6)\cdot(92$ nm x 10^{-7} cm/nm$)^3 = 4.29$ x 10^{-16} g/particle.
    $M = N_{avo} \cdot m = 6.022$ x 10^{+23} particles/mol $\cdot 4.29$ x 10^{-16} g/particle $= 258$ MDa.

    Remember a latex colloid is not a single molecule and its pseudo molecular weight has no meaning.

2. Estimate M using the closest power law in Table I-4. Clearly, dendrimers with an exponent of 0.30 is closest to an exponent of 1/3 expected for a sphere.

    So, try $46 = 0.08605 \cdot M^{0.30}$. M = 1,218 MDa, far from the solid sphere approach where M = 258 MDa.

    But, for a sphere, the exponent should be 1/3. Could such a small difference in exponent from 0.30 to 0.333 make such a significant difference?

    Now try $46 = 0.08605 \cdot M^{1/3}$. M = 153 MDa. Yes, relatively minor differences in exponents in MHS equations do make a significant difference. Interestingly, 153 MDa is much closer to 258 MDa than 1,240 MDa is. But, of course, this is not a single molecule.

    Summary: MHS equations are fun but given the large variation with end results with minor changes in the exponent, take care when working with such equations. Always better to make a measurement if instrumentation is available.

Chapter II: Static Light Scattering: M_w, R_g, A_2

II.1 Introduction to Light Scattering

I quote from something written 27 years ago[9]

"It is well known that the existence of light is mentioned at the beginning of the Bible. Less well known is a footnote in the bibliography to the following effects. Let everything scatter light. Let the relationship between particle size and scattering angle change dramatically over the size range of interest. And let ill-conditioned Laplace transforms run rampant over the whole field."

Anything with polarizable electrons scatters light, so atoms, molecules, and particles of all types do it. It was only a couple of years after Maxwell established light as a transverse electromagnetic wave that oscillates in space and time that Lord Rayleigh (J.W. Strutt) published his first articles showing how small particles scatter light[10]. The incident light's oscillating electric field causes polarizable electrons in the particle (or molecule) to oscillate and produce a scattered electric field proportional to polarization. The polarization is proportional to the number of electrons and so to the volume of the atom, molecule, or particle (an over-simplification, but one that illustrates the point). Since the volume is proportional to the size cubed, and since the scattered light intensity is proportional to the square of the scattered electric field, Rayleigh established the famous size raised-to-the-6th-power dependence for scattered light. He was also able to show the $1/\lambda^4$-dependence that accounts for blue skies and red horizons. Note that all this occurred before the full understanding of electrons. The

---

[9] B. B. Weiner, "Twenty-Seven Years of QELS: A Review of the Advantages and Disadvantages of Particle Sizing with QELS", editors N.G. Stanley-Wood and R.G. Lines, The Royal Chemical Society of Chemistry, Redwood Press Limited, Melksham, Wiltshire, 1992, Special Publication No. 102. ISBN 0-85186-487-2

[10] John Strutt (1871) "On the light from the sky, its polarization and colour," *Philosophical Magazine*, series 4, vol.41, pages 107–120, 274–279. John Strutt (1871) "On the scattering of light by small particles," *Philosophical Magazine*, series 4, vol. 41, pages 447–454.

existence of small, polarizable, charged entities was all that was necessary for Rayleigh's interpretation of Maxwell's equations.

Maxwell's equations showed that an electric field, a vector (**bold**), is a function of position (the vector **r**), time t, and the angular frequency of the light $\omega_o = 2\pi\nu_o$, where ν_o, the frequency, is related to the speed of light c and the wavelength in vacuum λ_o by $\nu_o \cdot \lambda_o = c$. The magnitude of the field is also a vector $\mathbf{E_o}$ and its direction is called the polarization of the field. (Not to be confused with polarizable particles.) See equation (II-1):

$$\mathbf{E_o}(\mathbf{r},t) = \mathbf{E_o} \exp(i\mathbf{q} \cdot \mathbf{r}) \exp(-i\omega_o t) \tag{II-1}$$

The wave vector **q** has a magnitude $q = 2\pi/\lambda = 2\pi n_o/\lambda_o$ where n_o is the refractive index of the medium in which the wave propagates.

In general, for irregularly shaped particles and molecules, the polarization **P** (a vector with variables $\mathbf{r}, t$) is related to the amplitude of the applied electric field by:

$$\mathbf{P}(\mathbf{r},t) = \underline{\mathbf{\alpha}}(\mathbf{r},t) \cdot \mathbf{E_o}(\mathbf{r},t) \tag{II-2}$$

Here $\underline{\mathbf{\alpha}}(\mathbf{r},t)$ is a tensor in general; however, for isotropic particles, it is a scalar. And its time dependence is very slow compared to that of $\mathbf{E_o}$ whose periodicity is on the order of 10^{-16} s.

For small particles, **P** reduces to the dipole moment induced by the electric field of the incident light. Maxwell's equations showed that an oscillating charge gives rise to another field, the scattered electric field, which is given by:

$$\mathbf{E_s}(\mathbf{r},t) = \frac{\frac{\partial^2 P}{\partial t^2}}{4\pi\varepsilon_o c^2 r} sin\varphi_z \tag{II-3}$$

Where c is the speed of light, r is the distance from the particle to the detector, ε_o is the permittivity of free space, and φ_z is the angle between $\mathbf{E_o}$ and the scattering plane. The scattering plane is defined as the plane encompassing the incident beam and the detected beam. In the most common configuration in modern times, with vertically polarized lasers, the scattering plane is horizontal and $\varphi_z = 90°$

Combining equations II-1, II-2 and II-3, and using $\omega_o = 2\pi\nu_o$ and $\nu_o \cdot \lambda_o = c$, results in:

$$\mathbf{E_s}(\mathbf{r},t) = \frac{-\alpha \cdot \pi \cdot \mathbf{E_o}(r,t)}{\varepsilon_o \cdot \lambda_o^2 \cdot r} \tag{II-4}$$

We can't measure these electric fields because they oscillate too fast for any detector to respond; instead, we measure the intensity, which is the dot (scalar) product of the electric field and its complex conjugate. Thus, intensity is a scalar and proportional to the square of the electric field. Therefore, the intensity scattered by one electric dipole, i_s, with I_o, the initial intensity, is given by:

$$i_s = \frac{\alpha^2 \cdot \pi^2 \cdot I_o}{\varepsilon_o{}^2 \cdot \lambda_o{}^4 \cdot r^2} \qquad \text{(II-5a)}$$

Grouping the variables differently, for reasons that will shortly become obvious, we have:

$$\frac{i_s \cdot r^2}{I_o} = \frac{\alpha^2 \cdot \pi^2}{\varepsilon_o{}^2 \cdot \lambda_o{}^4} \qquad \text{(II-5b)}$$

The polarizability is given by the Clausius-Mosotti equation found in many physical chemistry textbooks:

$$\alpha = \frac{3 \cdot M_p \cdot \varepsilon_o}{N_{avo} \cdot \rho_p} \cdot \left(\frac{n_p^2 - n_o^2}{n_p^2 + 2n_o^2} \right) \qquad \text{(II-6)}$$

Here M_p is molecular weight (or particle mass), ρ_p is the particle (or molecule) density and n is a refractive index: n_p and n_o are the refractive indexes of macromolecule (or particle) and the media in which it is dissolved or suspended. (For now, the scattering from the media is ignored.) The following substitutions are useful:

$$\frac{M_p}{N_{avo}\rho_p} = \frac{m_p}{m_p/v_p} = v_p = \frac{4}{3}\pi \cdot R_p^3 \qquad \text{(II-7)}$$

Where m_p and v_p are the mass and volume, respectively, of the scattering particle and, assuming a sphere, R_p is the radius. Substituting equations II-7 into II-6 and the result into II-5b, results in this equation:

$$\frac{i_s \cdot r^2}{i_o} = \frac{16 \cdot \pi^4 \cdot R_p^6}{\lambda_o{}^4} \cdot \left(\frac{n_p^2 - n_o^2}{n_p^2 + 2n_o^2} \right)^2 \qquad \text{(II-8)}$$

This equation shows that the scattered intensity has the following key features:

It varies with size raised to the sixth power. A 20 nm particle scatters 64 times that of a 10 nm particle. It follows that in a sea of small particles, the large particles dominate the scattering. If the large particles are not of interest—think dust (large and rare) or a few aggregates or dimers, etc.—they can cause problems in analyzing the signal.

Blue light (shorter wavelength) scatters more than red light (longer wavelength) by the ratio of the 4th power of the wavelengths: (632.8 nm/488 nm)[4] = 2.8, where 632.8 nm is the wavelength for a HeNe laser and 488 nm is the wavelength of the blue line of an Argon-ion laser. This effect explains why the sky is blue and the horizon is red: particles and gas molecules in air scatter the blue from the sun's rays preferentially, leaving the reds. Red penetrates more since it is scattered less. Thus, emergency and stop lights are generally red.

If the refractive index of the particle is the same as that of the media in which it is located, n_p = n_o, then there is no scattering at all. One way to achieve this is to dissolve synthetic polymers in a mixture of organic liquids to match the refractive indexes.

If there are N independent particles (far apart) in the observed scattering volume V_{obs} (defined as the intersection between the incident and detected light), we can add up the scattering intensities since there is no interference. The total scattered intensity I_s = $N \cdot i_s$. This introduces a concentration term, N/V_{obs}, and serves to define the Rayleigh Ratio as follows:

$$Rayleigh\ Ratio, R = \frac{I_s \cdot r^2}{I_o \cdot V_{obs}} = \frac{16 \cdot \pi^4 \cdot N \cdot R_p^6}{\lambda_o{}^4 \cdot V_{obs}} \cdot \left(\frac{n_p^2 - n_o^2}{n_p^2 + 2n_o^2}\right)^2 \tag{II-9}$$

The Rayleigh Ratio has units of inverse length, so it is not really a ratio, but the name has stuck. It consists of measured intensities and the geometry of the experimental setup (r, V_{obs}). It groups the instrument and measured intensities on one side of the equation and particle (or molecular) properties on the other.

Rayleigh's original equation was for gas molecules in vacuum, where n_o= 1 and n_p = $n_g \approx 1$. In this case, equation II-9 can be written as follows:

$$Rayleigh\ Ratio, R = \frac{I_s \cdot r^2}{I_o \cdot V_{obs}} = \frac{4 \cdot \pi^2 \cdot M_g}{\lambda_o{}^4 \cdot N_{avo} \cdot \rho_g} \cdot (n_g - 1)^2 \tag{II-10}$$

Equations II-9 and II-10 are for small particles or molecules far apart with insignificant scattering from the background. Consider a pure liquid. The molecules are close together. The scattered fields when summed would cancel out (destructive interference of electric field waves). Based on this, there should be no light scattering from a pure liquid. But there is. Pass a laser beam through toluene and the bright, sharp line is visual proof of scattering. From where does it come? From fluctuations in thermodynamic variables (P, T, c_p or S, V, c_p) due to random thermal motion. Replacing the square of the polarizability by the mean square fluctuation in the polarizability leads to scattering from pure liquids. See below for details.

II.2 Fluctuation Theory

The idea that scattering in pure liquids was not from individual polarizable molecules but from fluctuations in polarizability was first explored by Smoluchowski and Einstein[11]. Starting with equation II-5b and replacing α^2, the square of the polarizability, with $\overline{(\delta\alpha)^2}$, the mean square fluctuation in the polarizability, the jumping off point for fluctuation theory in light scattering becomes:

$$\frac{i_s \cdot r^2}{I_o} = \frac{\overline{(\delta\alpha)^2} \cdot \pi^2}{\varepsilon_o{}^2 \cdot \lambda^4} \tag{II-11}$$

---

[11] Smoluchowski, M., *Ann. Phys.*, **25**, 205 (1908); Einstein, A., *Ann. Phys.*, **33**, 1275 (1910).

In this case, since the light is in the medium, λ and not λ_o is used, where $\lambda = \lambda_o/n_o$.

For solutions of macromolecules in a liquid there is a bigger source of fluctuations: concentration fluctuations. These ideas are developed in greater detail in other sources.[12,13] We choose temperature (T), pressure (P), and macromolecular concentration (c_p) as the independent variables. Other choices are sometimes used such as entropy, volume (alternatively density) and concentration. The fluctuation in polarizability can be expressed as:

$$\delta\alpha = \left(\frac{\partial\alpha}{\partial P}\right)_{T,c_p} \delta P + \left(\frac{\partial\alpha}{\partial T}\right)_{P,c_p} \delta T + \left(\frac{\partial\alpha}{\partial c_p}\right)_{P,T} \delta c_p \tag{II-12}$$

The first two terms are the same for the solvent as they are for the solution. They give rise to solvent scattering that would otherwise have been zero. The last term gives rise to the excess scattering from the dissolved macromolecules. When the solvent scattering is subtracted from the total scattering, it is the excess that is left. In reference 12, section 5.3b, it is shown in detail how to evaluate $\left(\frac{\partial\alpha}{\partial c_p}\right)_{P,T}$ and δc_p. The end result is Eq. (D) in Table II-1.

Eq.(A)	$\dfrac{K \cdot c_p}{\Delta R_\theta} = \dfrac{1}{M_w}$	Small, noninteracting macromolecules
Eq.(B)	$\dfrac{K \cdot c_p}{\Delta R_\theta} = \dfrac{1}{M_w \cdot P_z(q)}$	Large, noninteracting macromolecules
Eq.(C)	$\dfrac{K \cdot c_p}{\Delta R_\theta} = \dfrac{1}{M_w} + 2 \cdot A_2 \cdot c_p$	Small, interacting macromolecules
Eq.(D)	$\dfrac{K \cdot c_p}{\Delta R_\theta} = \dfrac{1}{M_w \cdot P_z(q)} + 2 \cdot A_2 \cdot c_p$	Large, interacting macromolecules Zimm's one-contact approximation

Table II-1: SLS equations for a variety of conditions.

[12] Paul C. Hiemenz and Raj Rajagopalan, Principles of Colloid and Surface Chemistry, Chapter 5 "Static and Dynamic Light Scattering and Other Radiation Scattering", 3rd Edition, Marcel Dekker, Inc. publisher, 1997. Tanford, C., "Physical Chemistry of Macromolecules", Chapter 5, 1967, John Wiley & Sons, publisher.

[13] Kenneth S. Schmitz, An Introduction to Dynamic Light Scattering by Macromolecules, Chapter 2 "Basic Concepts of Light Scattering", Academic Press, Inc. publisher, 1990.

Where $P_z(q)$ has been added to take care of the cases where the particle or macromolecule is not very much smaller than the wavelength of light in the medium. In that case, the scattering from various parts of the particle or molecule must be added up to get the total and this depends on the scattering angle. Definitions of the various symbols are given below.

Here, $K = \dfrac{4 \cdot \pi^2 n_o^2 \left(\frac{dn}{dc_p}\right)^2}{N_{avo} \lambda_o^4}$, often given the symbol H, c_p is the macromolecular or polymer

concentration, ΔR_θ is the *excess* Rayleigh Ratio (the difference between $R_{solution}$ and $R_{solvent}$) obtained from the *excess* scattered intensity (the difference between $I_{s, \, solution}$ and $I_{s, \, solvent}$) at each scattering angle θ. (NOTE: ΔR_θ is sometimes denoted as R_θ but it is misleading. The scattering from the liquid must always be subtracted, unless solvent scattering is negligible.

M_w is the weight (mass) average molecular weight of the distribution of molecular weights (distribution of polymer chain lengths, for example). P_z is the z-average[14] particle scattering function and it is a function of θ through the definition of the scattering wave vector

$q = \dfrac{4\pi n_o}{\lambda_o} sin\left(\dfrac{\theta}{2}\right).$

A_2 is the second virial coefficient, and it is a measure of the difference between solvent-solvent, solute-solute (the macromolecules) and solvent-solute interactions. When $A_2 > 0$, true, single phase solutions are formed; when $A_2 < 0$, no solution is formed (two phases exist); and when $A_2 = 0$, which can happen when excluded volume effects are exactly equal to long-range attraction effects, the name *theta condition* is used to describe the macromolecular solution. This occurs for certain macromolecule-solvent pairs at a special temperature called the *theta* temperature.

N_{avo} is Avogadro's Number. More about these variables is discussed below, but it is worth mentioning when $c_p \to 0$ and either $\theta \to 0$ or $R_g/\lambda_o \to 0$, then $P_z \to 1$. And Eq.(A), Eq.(B), and Eq.(C) can derived from Eq.(D). The similarity between Eq.(II-9) and Eq.(A) should be noted. Both have the λ_o^4 dependence. Both are related to the size raised to the sixth power: $c_p M_w/N_{avo} \propto m_p^2 \propto v_p^2 \propto R_p^6$. Both show the same square dependence on the refractive index difference.

Whether it is scattering from particles or from macromolecules, the same dependencies arise.

The history of light scattering is replete with famous names and important discoveries that make it fundamental on theoretical grounds. See Table II-2.

---

[14] Z-averaging uses the intensity as the weighting function.

Light Scattering History

1871: Rayleigh Scattering Theory

1910: Einstein Fluctuation Theory

1944: Debye Theory for Polymer Solutions (SLS)

1948: Zimm Plots for M_w, R_g & A_2

1964: Cummins, Quasi Elastic Light Scattering (DLS)

1970s: Low Angle SLS for GPC/SEC

1980s: Ware, Flygare, Uzgiris (ELS)

1990s: Vincent, Miller, Schätzel, Tscharnuter (PALS)

Table II-2: Milestones in Light Scattering Theory & Instrumentation.

The first five establish various fundamental aspects of the theory. The last three are important in the progress of light scattering instrumentation.

II.3 The Basic Equation for Molecular Weight, Radius of Gyration & Second Virial Coefficient

In the limits of low concentration and scattering angle, the data obtained from SLS measurements for a dilute solution homopolymer in a single solvent can be written as [Eq.(D)]:

$$K \cdot c_p / \Delta R(\theta, c_p) = 1/[M_w \cdot P_z(q)] + 2A_2 \cdot c_p \tag{II-13}$$

It is useful to explore the parameters in greater detail. For vertical polarization, as before, the optical constant K is given by:

$$K = 4\pi^2 \cdot n_o^2 \cdot (dn/dc_p)^2 / (N_{avo} \cdot \lambda_o^4), \text{ where} \tag{II-14}$$

n_o is the refractive index of the solvent. Values from 1.30 to 1.60 are typical. For water, it is 1.331; for toluene, it is 1.497.

dn/dc_p is the specific refractive index increment of the solution. Values from 0.05 to 0.30 cm^3/g are typical. See Appendix SLS1 for details.

c_p is the polymer concentration. Values from 0.01 to 10 mg/cm^3 are typical. The lower concentration range is more common when measuring high molecular weight polymers, above 1 M Daltons, for example; the upper concentration range is more common when measuring low molecular weight polymers, below 100 K Daltons, for example.

N_{avo} is Avogadro's Number, 6.022×10^{23} mol^{-1}.

λ_o is the wavelength in vacuum of the laser. For a HeNe laser it is 632.8nm, and for an Argon-ion laser it is either 514.5nm (green) or 488.0nm (blue). For frequency doubled, solid state lasers it is 532 nm, and for diode lasers in the red it varies from 637 to 680 nm. The classical Hg arc lamp lines are 436 nm and 546 nm. (Old literature values.)

θ is the scattering angle. It varies from a few to nearly 180°, depending on the instrument and what is being measured.

$\Delta R(\theta, c_p)$ is the excess Rayleigh Factor (also called Rayleigh Ratio) in units of cm^{-1} calculated at different angles and concentrations. Excess means it is the <u>difference</u> between the value for the solution and the value for the solvent. R is proportional to the scattered light intensity; the proportionality constant is determined by calibration against a substance (typically toluene) with a known Rayleigh Ratio. See Appendix SLS2 for details.

M_w is the weight-average molecular weight (molar mass). Values in the range from a few hundred to tens of millions g/mol (Daltons) are measureable, more commonly 10^3 to 10^6.

$P_z(q)$ is the z-average particle scattering function or particle structure function. (In more concentrated and strongly interacting systems a solution structure function is also important.)

$q = (4\pi n_o / \lambda_o) \cdot \sin(\theta/2)$ is the magnitude of the scattering wave vector. See Appendix SLS3.

A_2 is the second virial coefficient. Values in the range of 0 to 10^{-4} cm$^3 \cdot$mol/g^2 are typical. For polydisperse systems, A_2 is also an average quantity; however, the average is not a simply-defined one like the number-, weight- or z-average.

The factor 4 in the optical constant K is a result of using incident light that is linearly polarized perpendicular to the horizontal scattering plane. This is the most common configuration and is typically referred to as vertical polarization. For un-polarized incident light, the factor is 2. It appears in old literature and textbooks before lasers were used.

The particle scattering function can be expanded in terms of q^2, and the first three terms are

$$P_z(q) = 1 - (R_g^2 \cdot q^2)/3 + O(q^4) \tag{II-15}$$

R_g is shorthand for $<R_g^2>_z^{0.5}$ the z-average, root-mean-square radius of gyration, a measure of the molecular size. R_g is almost universally called the radius of gyration, even though that is a misnomer. It is a root-mean-square value. In fact, even for a monodisperse, random-coil polymer, where the z-average equals all other average values, R_g is still an average value.

A random coil polymer takes up many configurations over time and R_g, even for a single molecule, is thus a temporally averaged value.

The coefficient of q^2 is independent of the particle shape. Coefficients of higher order terms do depend on shape. Therefore, the radius of gyration (more properly, the z-average, RMS radius) is a primary parameter obtained from light scattering: primary in the sense that no model for the shape is assumed. (Contrast with R_H from DLS where spherical shape is assumed.)

II.4 The Zimm Plot

Combining equations II-13 and II-15 and expanding $1/P_z(q)$ to first order in q^2 with a Taylor series, gives the following:

$$H \cdot c / \Delta R(\theta,c) = [1/M_w] \cdot [1 + (R_g^2 \cdot q^2)/3] + 2A_2c \tag{II-16}$$

This equation is the basis for the famous Zimm Plot. Developed in 1948 by Zimm[15], scattered light intensities are measured at several angles for each of several solution concentrations and for the pure solvent at each angle. Subtracting the solvent scattering from the solution value yields the excess intensity scattered by the polymer. A calibration constant is used to calculate excess Rayleigh Factors from measured intensities.

A calibration constant is obtained by measuring the intensity of light scattered from a standard. It is assumed that $R_c = k_c \cdot I_c$, where k_c is the calibration constant and R_c is the known Rayleigh ratio. R_c is known for toluene and benzene at the laser wavelengths. Rayleigh ratios scale primarily inversely with λ^4 and to a lesser extent with the wavelength dependence of the refractive index and depolarization ratio. Such a calibration is only good if the flare light at the calibration angle, typically $90°$, is insignificant. See Appendix SLS2 for a discussion of Rayleigh ratios for calibration standards (pure liquids) including tables of values at different wavelengths.

Breaking down the basic equation into a double extrapolation, one at fixed angle as concentration goes to zero, and one at fixed concentration as angle goes to zero, yields two straight lines. Each has the same intercept, $1/M_W$. And each has a slope: The slope of the angular dependence (q^2), after a little manipulation, yields R_g; and the slope of the concentration line yields $2A_2$. This is shown in equations II-17a and II-17b, respectively:

$$\frac{Kc}{\Delta R(\theta,c \to 0)} = \frac{1}{M_W} + \frac{R_g^2 q^2}{3M_W} \tag{II-17a}$$

---

[15] Zimm, B.H., J. Chem. Phys. **16**, 1093 (1948).

$$\frac{K \cdot c_p}{\Delta R_\theta} = \frac{1}{M_w} + 2 \cdot A_2 \cdot c_p \qquad \text{(II-17b)}$$

If the left-hand side of equation II-16 is plotted against $\sin^2(\theta/2) + k_p \cdot c$, where k_p is an arbitrary plotting constant chosen purely for convenience in viewing the results, a grid-like-plot results. Two sets of, typically, parallel lines make up the grid. One set consists of angular measurements at each concentration, and one set consists of concentration measurements at each angle. Extrapolating the angular measurements to zero-angle for each concentration yields a straight line in c. The slope of this line yields $2A_2$. Extrapolating the concentration measurements to zero concentration for each angle yields a straight line in $\sin^2(\theta/2)$. The slope divided by the intercept of this line yields R_g. Again, refer to equation II-17a.

The intercept of both extrapolated lines—the double extrapolation to zero angle and zero concentration—yields M_w, the weight-average molecular weight of the polymer in solution. It is this parameter of a polymer distribution that is most often of interest. And light scattering offers an absolute* method for its measurement. Strictly speaking, this is a misnomer for SLS.

[*An absolute measurement does not depend on any calibration factor. While the old Chromatix KMX-6 (not produced for decades) did not require calibration, most other SLS instruments do. However, once calibrated, changing polymer solutions does not require a recalibration. And the calibration depends on the properties of toluene, for example, a solvent that can be obtained as HPLC-grade at modest cost. This is in sharp contrast to column calibration using GPC/SEC.]

The value of dn/dc and several other parameters in equation II-14 as well the calibration constant must be known in advance to calculate M_w and A_2 from a Zimm Plot. This is also true for the Debye and Berry Plots described below. Interestingly, however, R_g can be calculated without knowing either dn/dc and without calibrating. These parameters cancel when calculating R_g. Thus, a z-average size is obtained, the z-average radius of gyration, with no assumption about shape, and without calibration or a value for dn/dc.

An example of a Zimm Plot is shown in Figure II-1.

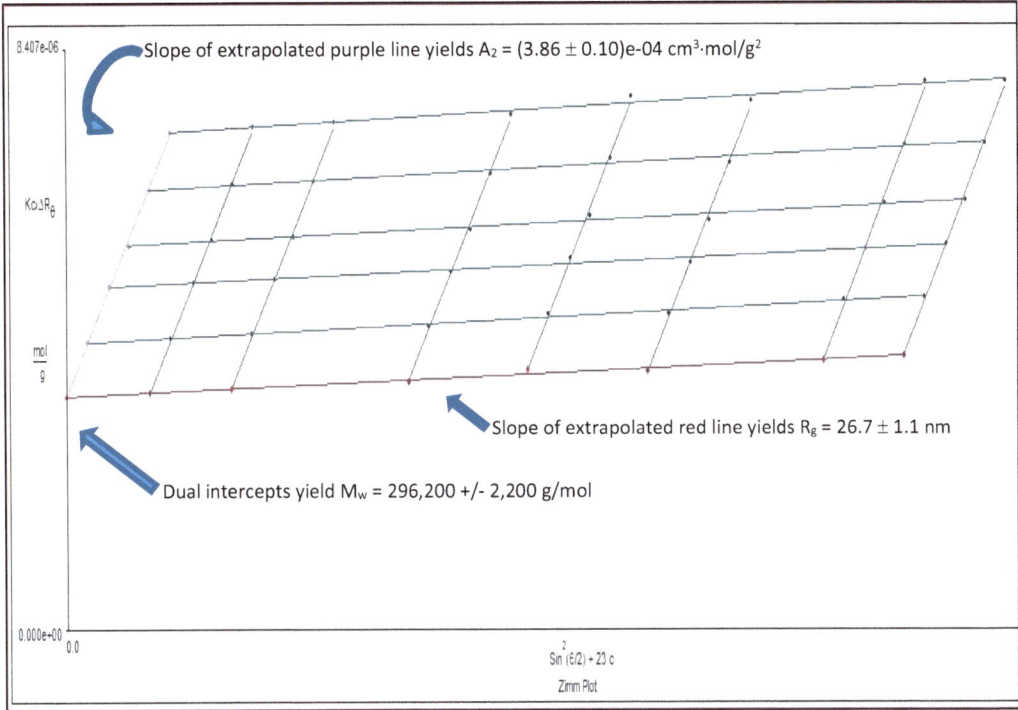

Figure II-1: A Zimm Plot using 5 concentrations and 7 angles.

Note: This is not a monodisperse sample. If it were, given M_w of 296.2 kDa for PS/Tol, R_g would be closer to 21.7 nm. See equation II-16. Thus, the sample has some polydispersity. Z-averages are always larger than weight-averages.

II.5 The Debye Plot

Rayleigh scattering is defined such that the size of the molecule or particle doing the scattering is negligible compared to the wavelength of light. In this case the angular term in equation II-16 is negligible, and it can be approximated by:

$$K \cdot c / \Delta R(\theta,c) = 1/M_w + 2A_2 \cdot c \qquad \text{(II-17b)}$$

This equation is the basis for the Debye Plot. At one angle intensities are measured as a function of concentration. The results are plotted according to equation II-15. A_2 is obtained from the slope of this plot; the intercept yields $1/M_w$. Typically, the angle used is 90°, but this is not necessary. This angle is convenient because flare light is often a minimum and because residual dust or micro gel contributions are usually minor.

Whenever the molecular weight of a random coil polymer is small, roughly 100 kDa and lower (thus, R_g < 10 nm), consider collecting data for a Debye Plot. The measurement time is shorter; the misinterpretation of possible downward curvature at low angles is avoided. Downward curvature may be due to dust, microgels, colloids, or flare. See Table II-3.

When to Use a Debye Plot

- Globular proteins since R_g normally < 10 nm

- Random coil polymers when M_W < 150-100 kDa

- Highly cross-linked polymers up to 1 MDa

- Dendrimers up to 100 MDa, if any exist

Table II-3: When to use a Debye Plot.

Starting with equation II-16, one gets to the Debye Plot, equation II-17b, when $R_g^2 q^2/3$ <<1. That occurs when $\theta \to 0$ (so $q \to 0$) or when $R_g \to 0$ (so a compact molecule). Comparing equations II-16 and II-17a, one can develop an exact equation for calculating the percent error in a calculated M_W vs. R_g and q (so θ, λ). See Table II-4.

For example, if you are willing to accept a 5% error in M_W, you can use a Debye plot at 90° with a red diode laser at 640 nm all the way up to a R_g of 21.1 nm. But if you work in backscatter with a HeNe laser (633 nm), you can't work above R_g of 15.1 nm if you want to keep the error at or below 5%. Lower angles and longer wavelengths are always better in keeping the error in M_W small when using a Debye Plot.

Limitations of Debye Plot

$$\%Error\ in\ M_W = 100 \cdot \frac{\dfrac{R_g^2 \cdot q^2}{3}}{1 + \dfrac{R_g^2 \cdot q^2}{3}}$$

% Error	R_g(640 nm, 90°)	R_g(633 nm, 173°)
1	9.4	6.6
2	13.3	9.4
5	21.1	15.1
10	29.8	21.9

Table II-4: Errors in M_w when a Debye Plot is incorrectly used.

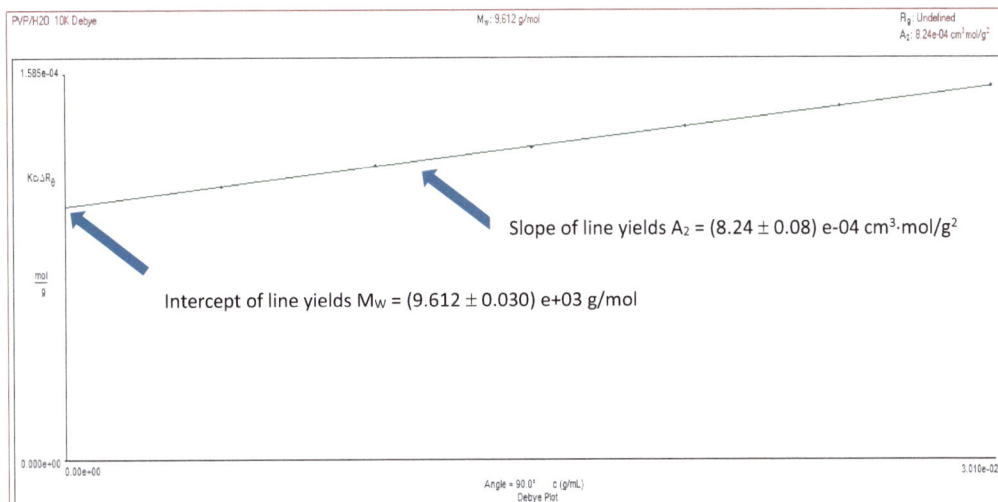

Figure II-2: A Debye Plot at 90°using 6 concentrations. The sample was a nominally 10 kDa poly(vinylpyrrolidone) in water.

II.6 The Berry Plot

The linear terms in c in equations II-16 and II-17b are the first in a polynomial expansion in c. These terms come from the expansion of the osmotic pressure in c. At higher concentrations or for strongly interacting particles (polyelectrolytes not screened by added free salt ions), they may be important. They are also important, for different reasons, for branched polymer systems. Whatever the reason, the Zimm Plot shows upward curvature, and it is no longer a simple set of straight lines.

The curved data may be fit to higher order terms in c and $\sin^2(\theta/2)$. Software may provide for fits of higher order in either or both variables. With few data points, it is always possible to fit the measured data quite well with a high-order polynomial. Nevertheless, the parameters calculated with high-order fits are often not meaningful. Whenever possible use a linear fit or try to linearize the data. The Berry Plot was developed for this purpose.

The right-hand side of Eq. (II-16) can be written as:

$$[1/M_w] \cdot [1+(R_g^2 \cdot q^2)/3] \cdot [1+2A_2 \cdot M_w c+3A_3 \cdot M_w \cdot c^2+ \dots] \qquad \text{(II-18)}$$

For hard spheres, it has been shown that $A_3 M_w = (5/8) \cdot (A_2 M_w)^2$, while for flexible coils in a theta solvent the term is zero. In general, it can be written as $g(A_2 \cdot M_w)^2$. When the 3rd order term is small but not negligible, an approximation for g might be $1/3$, the approximate average between 0 and $5/8$ (.625). With $g = 1/3$,

29

$$[1 + 2A_2 \cdot M_w \cdot c + (A_2 \cdot M_w)^2 \cdot c^2 + ...] \cong [1 + A_2 \cdot M_w \cdot c]^2 \qquad \text{(II-19)}$$

Substituting into Eq. (II-16), the result is given by:

$$H \cdot c / \Delta R(\theta,c) \cong [1/M_w] \cdot [1 + (R_g^2 \cdot q^2)/3] \cdot [1 + A_2 \cdot M_w \cdot c]^2 \qquad \text{(II-20)}$$

This equation can be made to yield two sets of lines by taking the square root. The result, linear in q^2 and c, is given by:

$$[H \cdot c / \Delta R(\theta,c)]^{0.5} = [1/M_w]^{0.5} \cdot [1 + (R_g^2 \cdot q^2)/6] \cdot [1 + A_2 \cdot M_w \cdot c] \qquad \text{(II-21)}$$

This equation is the basis of the Berry Plot. The square root of $K \cdot c / \Delta R$ is plotted against $\sin^2(\theta/2) + k_p \cdot c$.

II.7 The 2nd Virial Coefficient

In the polymer and light scattering literature, one often finds an alternate designation for the second virial coefficient, Γ. The value of Γ typically varies from about 10 to 1,000 cm^3/g, and it is a measure of molecular volume. The symbol Γ is often used in the discussion of viscosity of polymer solutions. The symbol Γ is related to A_2 as follows:

$$A_2 = \Gamma/M \qquad \text{(II-22)}$$

For example, assume a polystyrene sample with $M_w = 100,000$ g/mol (Da) and $A_2 = 1 \times 10^{-4}$ $cm^3 \cdot mol/g^2$. Then $\Gamma = 10$ cm^3/g. The mass of one polymer molecule is $100,000/6.022 \times 10^{23} = 1.66 \times 10^{-19}$ g/molecule. Since Γ is a measure of molecular volume, 1.66×10^{-19} g/molecule x 10 $cm^3/g = 1.66 \times 10^{-18}$ cm^3/molecule = 1.66×10^3 nm^3/molecule. Thus, Γ has represents a molecular volume. Set volume equal to $(4/3)\pi R^3$, and the calculated radius is 7.3 nm. But it is a little hard to say it is the same as R_H obtained from DLS or R_H calculated from R_g. Still, it is close.

II.8 SLS as an Absolute Molecular Weight Detector for GPC/SEC

Gel Permeation Chromatography, also known as Size Exclusion Chromatography, utilizes columns to separate molecules of distinct size as they flow in an eluent. Classically, the system was calibrated with standard polymers of known molecular weight under the assumption that they interacted with the column the same way as the unknown. Another method, dubbed, prematurely, universal calibration, uses a differential viscometer as the detection method. Many nonpolar polymers in many nonpolar solvents exhibited a common signal vs. elution volume curve, and this gave rise to the "universal" moniker. Many polar macromolecules (proteins, for example) in polar solvents (water, for example) do not follow the universal curve. But they all scatter light.

The system is flowing and the concentration at any given slice (typically of 1 second duration) is on the order of 10-50 μg/mL; whereas, for batch mode determinations in Zimm Plots, the concentration is in the range of 1-10 mg/mL. What does that mean for the term $2 \cdot A_2 \cdot c$ compared to $1/M_W$? With a range of 10^{-3} to 10^{-5} for A_2, the range for $2 \cdot A_2 \cdot c$ is 2×10^{-10} mol/g to 10^{-7} mol/g. Compare that to $1/M_W$. Only for molecular weights above 1 MDa would the term $2 \cdot A_2 \cdot c$ be significant. Therefore, the basic equation reduces to this:

$$H \cdot c / \Delta R(\theta, c) = 1/M_w + (R_g^2 \cdot q^2)/3 \cdot M_W \qquad \text{(II-23)}$$

This is analogous to the Debye plot, but here it is M_w and R_g that are determined. It is Eq. (B) and equation II-17b. Thus, adding an SLS device to the end of the detector train on an SEC/GPC device leads to molecular weight and radius of gyration.

II.9 Corrections and Measurement Considerations

Before fitting the data and accepting the results, consider the following:

II.9.1 Intensity Measurements

For a complete discussion of the actual measurement of intensity using photon counting and statistics, see Appendix SLS4.

II.9.2 Depolarization Corrections

For a complete discussion of how to treat measurements on anisotropic molecules (rigid rods, for example), see Appendix SLS5.

II.9.3 Measurement Considerations

Real samples and real measurements may not always fit exactly the theory just presented above. As the angle is decreased, the intensity is often anomalously large: spikes due to dust, scratched cells, micro gels, or other supramolecular entities, and, of course, flare-light at the cell-solution and cell-air interfaces. These effects cause a downward curvature in the Zimm and Berry Plots. One should consider whether this downward curvature is real. If real, it will lead to lower intercepts and greater slopes, resulting in large M_w and R_g values. If unreal, it will lead to anomalously large values.

Such curvature is easily fit to higher order polynomials; yet, caution is advised for any fit higher than quadratic. Though the fit will certainly improve as the order of the polynomial is increased, one must use judgment in accepting these results. Are they repeatable results? If not, then dust, micro gels, and other supramolecular entities (colloids, perhaps) are most likely the problem. Were spikes in the repeated intensity measurements obvious? Then dust or its equivalent is most likely the cause. Clean up the sample before repeating the measurement. If the results are repeatable, do they make sense? If not, perhaps flare light is the problem; perhaps the calibration was not done with a clean liquid; perhaps the polymer concentrations are not accurate; perhaps the value of dn/dc is wrong.

Interestingly, columns used in GPC/SEC also act as filters against dust, microgels and other supramolecular entities. Thus, this added benefit may help in reducing anomalous downward curvature and erroneously larger values.

II.10 Generalized Intensity Scattering: Structure

Total scattering includes interparticle and intraparticle effects, S(q) and P(q), respectively. Analyzing P(q) will lead to particle structure; analyzing S(q) will lead to liquid structure (complex fluids). Total scattering can be broken down into four factors, the first of which is a calibration constant, k_c:

Total Scattering = k_c x contrast x concentration x structure factor(s)

For SLS, the contrast factor for polymers and proteins in solution is proportional to $(dn/dc)^2$ and for particle suspensions it is $\left[\frac{(n_p - n_l)}{(n_p + 2n_l)} \right]^2$. If the refractive index differences are small enough, the scattering is too weak to be properly measured. If the surface chemistry allows it, find a liquid with a greater index difference.

For SLS, the concentration factor can be written in equivalent ways. For molecular weight determination, there is the familiar c·M term. For particles in suspension there are the familiar $N \cdot d^6$ or $c \cdot d^3$ terms.

For SLS, the structure factor is proportional to the product of P(q)·S(q). As c → 0, liquid structure disappears and S(q) → 1. That leaves P(q), which has many names: the particle scattering function, the intraparticle structure factor and the particle form factor. It is also called the Mie scattering function when the particles are spheres. Its angular dependence leads to particle size (R_g).

II.11 Intraparticle Structure: Particle Sizing, P(q)

In general, as shown by theory, P(q) approaches 1 when θ → 0 or R_g/λ → 0. This is embodied in the expression P(q) → 1 − $(R_g \cdot q)^2/3$ which is <u>independent of shape</u>. It is true for light, x-ray and neutron scattering and is derived from scattering equations with the definition of R_g given by:

$$R_g^2 = \Sigma_i\, m_i\, r_i^2 / \Sigma_i\, m_i \qquad \text{(II-24)}$$

Here m_i is the ith mass element of the scattering particle and r_i is the distance from the center of gravity to the ith element. The interesting thing about this definition is that it is independent of the shape of the particle, although to interpret the measured R_g further requires a model. For a sphere, $R_g^2 = 3/5\ R^2$ where R is the radius of the sphere, and for long, thin rods $R_g^2 = L^2/12$ where L is the length of the rod. See Chapter I. For other shapes consult specialized texts.[16]

Please note that R_g is also and more properly called the root-mean-square radius. Unfortunately, most textbooks and publications use the term radius of gyration or root-mean radius of gyration, and it is difficult to break the habit. [Witness: molecular weight vs. relative molar mass.] In addition, for a polydisperse system, $R_g = <R_g^2>_z^{1/2}$, the z-average, root-mean-square radius. Z-averaging is a result of intensity weighting and has some interesting consequences which will not be explored further. Contrast this with the weight-averaging of the molecular weight also obtained from light scattering.

II.12 RGD Theory

If the particle is not too large, a result that spans from where Rayleigh theory leaves off (starting around 25 nm) and full Mie theory starts up (300-350 nm), is the Rayleigh-Gan-Debye theory for spheres abbreviated as RGD. In this case, theory shows the following:

$$P(q) = \left\{ \left[\frac{3}{(qR)^3} \right] \cdot \left[\sin(qR) - qR\cos(qR) \right] \right\}^2 \tag{II-25}$$

Note that, like Rayleigh theory, this does not require the particle refractive index. Also, you can check that, in the limit as $q \cdot R \to 0$, the expression also goes to $P(q) \to 1 - (R_g \cdot q)^2/3$ where, for spheres $R_g^2 = 3/5\ R^2$.

RGD theory applies up to about half the wavelength of visible light, so maybe 300-350 nm in diameter and for the real part of the refractive index not much greater than 1.6. When plotted, the RGD equation exhibits minima and oscillations much like full Mie theory. Assuming the particles are spherical, it is still useful in calculations over which it applies.

II.13 Guinier Plot

Guinier[17] proposed the following simple variation in the excess scattered intensity with scattering angle for large and globular structures:

$$I_{ex}(q) = C \cdot \exp(-R_g^2 q^2 / 3) \tag{II-26}$$

---

[16] Burchard, Static and Dynamic Light Scattering from Branched Polymers and Biopolymers, in Advances in Polymer Science, Vol. 48, Light Scattering from Polymers, table 3, pages 74-75, Springer-Verlag, New York, 1983.
[17] Guinier & Fournet, Small Angle Scattering of X-rays, Wiley Interscience, 1955

Plotting $\ln(I_{ex})^{-1}$ vs. q^2 yields a straight line when a Guinier plot is appropriate. The slope is $R_g^2/3$ from which the radius of gyration can be determined. A Guinier plot is also useful, indeed was first suggested, <u>for small angle x-ray scattering</u>, because the wavelengths used are sub nanometer. Therefore, "large" means sizes of a few nanometers, a perfect range for globular proteins. Still, small angle x-ray scattering is expensive and sizing of globular proteins using DLS to obtain R_H is easier. See Figure III-3 for a Guinier Plot.

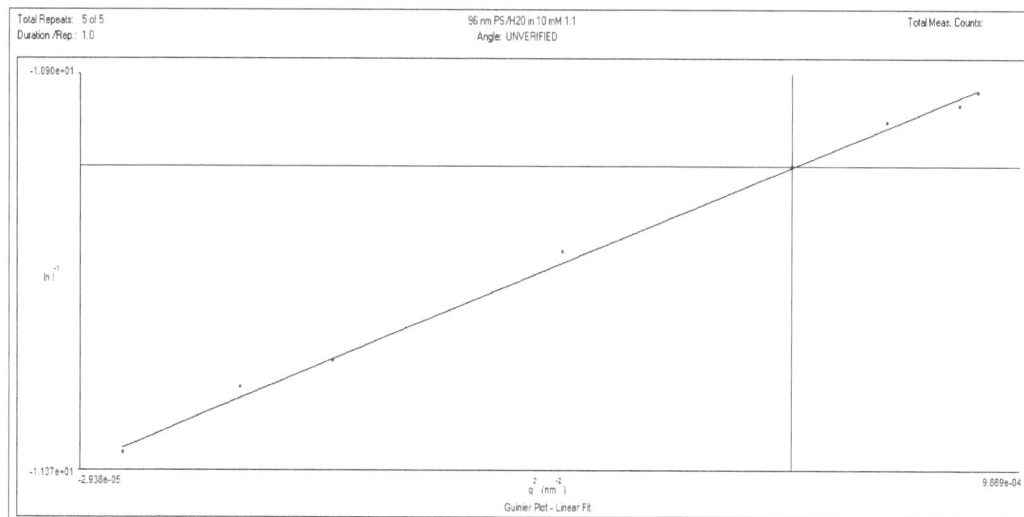

Figure III-3: A Guinier Plot using 8 angles from 15° to 155°.

The slope equals $R_g^2/3$ from which R_g is calculated as 37.0 +/- 0.1 nm. From DLS R_H = 96.0 +/- 2.0 nm. Thus, $R_g/R_H = 37.0/48.0 = 0.771$, quite close to $(3/5)^{1/2} = 0.775$, consistent with the fact these are latex spheres.

II.14 Partial Zimm and Berry Plots

For smaller particles and molecules, the partial Zimm and partial Berry plots are useful in the determination of R_g. The word "partial" used here signifies that only the angular variation is considered, not the concentration variation that is required for the determination of the weight-average molecular weight of polymers in dilute solution. Here, these plots are exclusively for determination of R_g.

The partial Zimm plot stems from the following approximate formula:

$$1/I_{ex}(q) = c \cdot \left(1 + \left(R_g \cdot q\right)^2/3\right) \tag{II-27}$$

Here R_g is determined from the slope and the intercept (c, a constant) of a plot of $1/I_{ex}(q)$ vs. q^2.

Zimm and partial Zimm plots are appropriate when the third virial coefficient is negligible. Virial coefficients describe the interaction between particles and are beyond the scope of this discussion. However, when the polymer concentration is too high or when the interactions are strong (polyelectrolytes at low salt concentration, for example), an approximation suggested by Berry may be used.[18]

The partial Berry plot follows from the following equation:

$$(1/I_{ex}(q))^{1/2} = c \cdot \left(1 + \left(R_g \cdot q\right)^2/6\right) \qquad \text{(II-28)}$$

Here R_g is determined from the slope and the intercept of a plot of $(1/I_{ex}(q))^{1/2}$ vs. q^2. Note the 1/6 instead of 1/3. This arises naturally when taking the square root of the term in parenthesis in the Zimm plot, assuming $(R_g \cdot q)^2/3$ is less than one.

Why do Zimm, Berry, and Guinier Plots look similar?

For Zimm and Berry plots, the answer lies in a Taylor expansion when $(R_g \cdot q)^2/3 < 1$. Partial Zimm plots start out as $1/P(q) = 1/[1 - (R_g \cdot q)^2/3]$ and become $1 + (R_g \cdot q)^2/3$. And partial Berry plots start out as $[1/P(q)]^{0.5} = \{1/[1 - (R_g \cdot q)^2/3]\}^{0.5}$ and become $1 + (R_g \cdot q)^2/6$. Guinier plots start out as $C \cdot \exp[-(R_g \cdot q)^2/3]$ and become $LnC^{-1} + (R_g \cdot q)^2/3$. Thus, all three produce straight lines vs q^2 from which R_g is calculated from the slope.

II.15 Kratky Plots

For branched polymers information on the extent of branching may be obtained from the shape of the Kratky plot, where $(q \cdot I_{ex}(q))^2$ vs. q is plotted. Highly branched polymers display a distinct maximum, and unbranched polymers do not. However, polydispersity also affects the shape of the plots, and the user should be careful in interpreting the shape of the plot without additional information.[19]

II.16 Fractals

Some interesting colloidal particles result from the aggregation of smaller subunits. The structures obtained display a variety of irregular shapes, some more open than others, including branching. The openness of the aggregates is described by a non-integer number called the fractal dimension, d_f: the larger the fractal dimension, the more compact the structure;

---

[18] Berry, J.Chem.Phys., 44, 4550 (1966)

[19] For more information consult Burchard, Macromolecules, 10, 919 (1977).

the lower the fractal number, the more open the structure. Spheres have a fractal dimension of 3.

The determination of the fractal dimension by scattering is appropriate when the size of the subunit, R_o (radius of an assumed spherical monomer), and the radius of gyration of the aggregate, R_g, satisfies the following criteria:

$$R_o < q^{-1} < R_g \qquad\qquad \text{(II-29)}$$

Think of a bunch of grapes clustered together. An individual grape might have a diameter of 1 cm so $R_o = 0.5$ cm, but the bunch might span 15 cm with perhaps an estimated R_g around 11 or 12 cm. For this case 0.5 cm < $q^{-1} = \lambda/[4\pi\cdot\sin(\theta/2)]$ < 11 cm and either the wavelength must be very long or the scattering angle very low. In any case, such an instrument doesn't exist, but the fractal dimension for macroscopic cases can be calculated geometrically.

Fractal dimensions also indicate how the aggregate was formed: limiting, ideal cases are diffusion-limited, cluster-cluster aggregation (DLCCA or DLCA) and reaction-limited, cluster-cluster aggregation (RLCCA or RLCA).[20]

When a determination of the fractal dimension is *appropriate*, the following equation is applied:

$$I_{ex} = c \cdot q^{-d_f} \qquad\qquad \text{(II-30)}$$

The fractal dimension is obtained as the slope of $\ln(I_{ex}(q))$ vs. $\ln(q)$. Ln-Ln plots are notorious for yielding seemingly straight lines. However, the reader is warned that such plots do not always yield straight lines over a sufficiently large Ln(q) space (one to two orders of magnitude) for the resulting slope to be usefully interpreted as the fractal dimension.

Equation II-30 is the limiting case of the more complete Fisher-Burford equation[21] when $R_g \cdot q >> 1$:

$$S(q) = S(0)/\left[1 + (R_g \cdot q)^2 \cdot \frac{2}{3d_f}\right]^{d_f/2} \qquad\qquad \text{(II-31)}$$

S(q) varies as q^{-d_f}. Furthermore, P(q), which is normally $1 + (3/5R_o^2\cdot q^2)/3 = 1 + R_o^2\cdot q^2/5$, goes to 1 because the condition for determination of d_f by scattering requires $R_o \cdot q < 1$.

---

[20] See Carpineti, Ferri and Giglio, Physical Review A, 42, 7347, (1990) for more information.

[21] Fisher,M.E., and Burford,R. J.(1967). Theory of Critical-Point Scattering and Correlations I. The Ising Model, Phys. Rev. A 156:583–622.

Thus, the excess intensity I_{ex}, which is proportional to $S(q) \cdot P(q)$ reduces to $S(q) \propto \boldsymbol{q}^{-d_f}$. Beyond molecular weight, static light scattering yields information on structure: size and shape of molecules and colloids; fractal dimensions of irregularly shaped particles; and structure in liquids, especially complex fluids. However, there are usually limitations (multiple scattering, for one) and the reader is well advised to consult the literature.

II.17 Beyond Light Scattering: X-ray and Neutron Scattering

In SLS the independent variable is q, the magnitude of the scattering wave vector. While light scattering is the main topic of this chapter, it is useful to expand the scope of the discussion to make the reader aware of its close relationship to other types of scattering: namely, small-angle x-ray scattering (SAXS); small angle neutron scattering (SANS); and wide-angle neutron scattering (WANS). In each case, a source of radiation is necessary (light, x-rays, neutrons) and the angular pattern of scattered intensity is measured. In each case, the scattering wave vector takes the same form $q = (4\pi/\lambda) \cdot \sin(\theta/2)$. The scattering wave vector defines the size range over which the type of scattering is useful. See Table II-5.

Method	Typical λ in nm	Range of q (nm^{-1})	Range of q^{-1} (nm)
SLS	500	$0.001 \rightarrow 0.04$	$25 \rightarrow 1{,}000$
SAXS	0.15	$0.02 \rightarrow 0.4$	$2.5 \rightarrow 50$
SANS	0.4	$0.007 \rightarrow 0.9$	$1.1 \rightarrow 140$
WANS	0.4	$10 \rightarrow 50$	$0.02 \rightarrow 0.1$

Table II-5: Different static scattering methods.

For example, the table suggests that the minimum size using light scattering is approximately 25 nm. Think of that as the diameter of the smallest particle detectable by SLS and that fits in nicely with the discussion about a minimum R_g in the range of 9 to 12 nm.

For proteins, quantum dots, and nanoparticles smaller than 25 nm, static light scattering is not going to produce size results. Compared to the wavelength, they are points. And points have no dimensions. Dynamic light scattering (DLS) on species that small is viable (see Chapter III) as is small angle x-ray and neutron scattering. The range 1 to 140 nm would be extraordinary useful in nanotechnology (nominally 1 to 100 nm). It is, unfortunately, a fact that neutron scattering can only be done at a few national labs around the world and that SAXS is much more expensive than DLS. A hidden advantage of neutron scattering is its immunity to multiple scattering which limits light scattering, SLS and DLS.

There are many related plots in light, x-ray, and neutron scattering: The x-axis can be q, q^2, Ln(q). The y-axis can be I_{ex}, $(I_{ex})^{-1}$, $Ln(I_{ex})^{-1}$, or $q^2 \cdot I_{ex}(q)$. There are other, more specialized plots.[22]

II.18 Interparticle Structure: The Static Structure Factor S(q)

In advanced texts on this subject it is the z-average particle scattering factor $P_z(q)$ that contains the variation of intensity with q after the volume correction [$sin(\theta)$] has been applied. $P_z(q)$ is also called the form factor or the intraparticle structure factor. It contains information on the structure or form of the particles. At higher concentrations, when interparticle interactions are important, there is another term called the static structure factor or the interparticle structure factor, S(q), which contains information on the distribution of the particles in the liquid. [There is also a dynamic structure factor, accessible using dynamic light scattering.] The intensity $I_{ex}(q)$ is, for randomly oriented, monodisperse particles, a product of $P_z(q)$ and S(q). For the dilute systems considered here, S(q) is constant. Again, for convenience, we refer here just to intensities. It is up to the user to account for the more complex cases.

Static Structure Factor, S(q)

- Contains information on interparticle interactions & structural details of aggregates: radial distribution functions, g(r), and fractal dimensions, d_f.

- S(q) → Constant, when q→ 0, that is θ→ 0 or when concentration → 0. And, therefore we ignore S(q) when doing measurements in dilute systems, focusing solely on $P_z(q)$ and what it might tell us about particle size.

- Example of S(q): Interparticle Interference Effects

$$S(q) = 1 + \frac{4\pi c_N}{q} \cdot \int_0^\infty [g(r) - 1] \cdot r \sin(q \cdot r) \cdot dr$$

Here c_N is the number concentration. Measurement of S(q) and deconvolution to determine g(r) yields information on the structure around a particle in the liquid. Generally, the concentration at which S(q) is measureable in light scattering is too high and multiple scattering masks the effects. With neutron scattering, multiple scattering doesn't occur and S(q) can be determined.

---

[22]See https://www.ncnr.nist.gov/staff/hammouda/distance_learning/chapter_22.pdf for a detailed discussion.

Chapter III: Dynamic Light Scattering, DLS

III.1 Introduction

Particles move. In a gas, they move very fast and their mean-free path is much longer than their size. In a liquid, they also move fast, but their mean-free path is much shorter than their size. This has consequences for their time-dependent motion and the time dependence of scattered light. It leads, in a liquid, to diffusion coefficients from which particle or molecular size is calculated.

Classical light scattering, CLS, more often called Static Light Scattering, SLS, averages the intensity of scattered light over timescales much longer than that which describes the motion of the particle. A one second averaging is more than enough to eliminate any time-dependent effects of the motion. Depending on size, such motion takes place on microsecond and millisecond timescales. Time-dependent LS has many names. See Table III-1.

**Time-Dependent Light Scattering
By Any Other Name**

- DLS, Dynamic Light Scattering
- PCS, Photon Correlation Spectroscopy
- QELS, Quasi-Elastic Light Scattering
- IFS, Intensity Fluctuation Spectroscopy

Table III-1: Different names for the same thing causes confusion for the novice.

Originally called IFS, intensity fluctuation spectroscopy, the technique was used to determine statistical properties of scattered light. It has been used to determine dynamic structure factors in complex fluids, microrheological properties of complex fluids, molecular weights of polymers and proteins in solution (via the Mark-Houwink-Sakurada equation); however, its largest application is for sizing dissolved polymers and proteins in solution and particles in suspension in the colloidal and nanoparticle size range.

Because the incident light has an insignificant effect on the particle's movement, this type of scattering is almost purely elastic: it is quasi-elastic. [Please do not refer to it as "quells", but rather Q.E.L.S., sounding out each letter separately.]

Initially, the power spectral density function of the intensity of scattered light, $P_I(\omega)$, was measured using a spectrum analyzer, but later the autocorrelation function of the intensity, $C_I(\tau)$, which is the Fourier transform of $P_I(\omega)$, was measured directly using a digital autocorrelator. This preserved the digital nature from photons exiting the laser to scattered photons from the particles to photons generating digital signals at square-law detectors (PMT's & APD's) that were correlated. This is many times more sensitive than the analog measurements of $P_I(\omega)$. As a result, for much of the 1970's, the technique was called photon correlation spectroscopy. [$P_I(\omega)$ re-enters the picture when discussing electrophoretic light scattering, ELS, and zeta potential.]

But by the early 1980's, the name dynamic light scattering, DLS, was catching hold. It makes a nice contrast with SLS and ELS, electrophoretic light scattering, or PALS, phase analysis light scattering. DLS is a name that reflects the dynamics of the particle as it diffuses. And that dynamics is embodied in the translational diffusion coefficient, D_T. [For rod-like particles there is also rotational diffusion. For flexible coil polymers, there is internal motion. These are special cases, but here we describe rigid, globular particle motion.] The several types of LS are summarized in Table III-2.

Type	Measure	Determine
Static SLS	Average Intensity	M_W, R_g, A_2, Distributions (GPC/SEC detector), Static Structure Factors
Dynamic DLS	Correlation Function	Average Size, Size Distribution, Relaxation Rates, Microrheological properties, Dynamic Structure Factors
Electrophoretic ELS	Doppler Shift	Electrophoretic Mobility, Average Zeta Potential, and Zeta Potential Distributions in Simple Cases
Phase Analysis PALS	Phase Shift	Electrophoretic Mobility, Zeta Potential in Simple and Difficult Cases (solvents, high salt, viscous samples), Solid Surface Zeta Potential

Table III-2: Several Types of Light Scattering. ELS is sometimes called LDE, laser Doppler electrophoresis, after the more general name LDV, laser Doppler velocimetry.

III.2 General References

Much of the information in this chapter can be seen in greater detail in one or more of the following references:

A. **Laser Light Scattering**, authors Charles S. Johnson Jr. and Don A. Gabriel, Dover Publications, Inc., New York. This monograph, originally published in 1981 by the CRC Press, was reissued in 1994 with corrections and added material. It is an excellent review of the earlier work including the frequency measurements before fast, digital correlators were available. Much of the emphasis is on biochemical applications, especially the early work from 1964 to 1980. A section on electrophoretic light scattering is also interesting.

B. Chapter 4, Data Analysis in Dynamic Light Scattering, Petr Štěpánek, in **Dynamic Light Scattering: The Method and Some Applications**, author Wyn Brown, Oxford Science Publications, Clarendon Press, Oxford, UK, 1993. This chapter is an excellent review with references for all the popular data analysis methods up to circa 1991. Other chapters in this book cover scattering from polymer solutions, both dilute and semi-dilute, scattering from gels and other complex fluids. It is an excellent, advanced text with many good references on these subjects.

C. **Dynamic Light Scattering by Macromolecules**, author Kenneth S. Schmitz, Academic Press, 1990. Despite its title, this book covers static light scattering, scattering from colloids and complex fluids as well as emphasizing the title material.

D. **Laser Light Scattering: Basic Principles and Practice, Second Edition**, author Benjamin Chu, Academic Press, 1991. An excellent, overall presentation it includes many of the earlier experimental designs, the overlap of x-ray, neutron, and light scattering. There is material devoted to using single mode fiber optics in SLS and DLS.

III.3 Brief Introduction to Intensity Power Spectral Density, $P_I(\omega)$, and Autocorrelation, $C_I(\tau)$, Functions

Time dependence of the scattered light intensity (subscript "I") can be determined in two ways: passing the current generated in a photomultiplier tube from light impinging on its photocathode to obtain the powder spectral density function, $P_I(\omega)$; or accumulating the autocorrelation function, $C_I(\tau)$, obtained from photons impinging on the PMT or on an avalanche photodiode (APD). Both techniques yield the same information. In fact, $P_I(\omega)$ and $C_I(\tau)$ are Fourier transform pairs related by the Wiener-Khinchine Theorem:

$$P_I(\omega) = \frac{1}{\pi} Re \int_0^\infty C_I(\tau)\, e^{-i\omega\tau} d\tau \qquad \text{(III-1)}$$

$$C_I(\tau) = \frac{1}{\pi} Re \int_0^\infty P_I(\omega)\, e^{i\omega\tau} d\tau \qquad \text{(III-2)}$$

Likewise, the power spectral density function of the scattered electric field, $P_E(\omega)$, and the autocorrelation function of the scattered electric field, $C_E(\tau)$, are also Fourier transform pairs.

The scattered electric field autocorrelation function $C_E(\tau)$ is also called the 1st-order autocorrelation function; and the scattered intensity autocorrelation function $C_I(\tau)$ is also called the 2nd-order autocorrelation function. It is the 1st-order autocorrelation function that is related to particle or molecular properties, but it is the 2nd-order autocorrelation function, or its transform pair $P_I(\omega)$, that are measured.

Note on Nomenclature

Over the years different authors have used different symbols. Early authors used the symbols we will employ in this chapter: τ for delay time; τ_r for decay (relaxation) time (see below); and superscripts enclosed in parentheses to designate first and second order autocorrelation functions such as $g^{(1)}$ and $G^{(2)}$. The lower-case g refers to normalized and the upper-case G refers to unnormalized autocorrelation functions. Thus, $G^{(2)}(\tau) = C_I(\tau)$, the measured, unnormalized intensity autocorrelation function. Later authors used t for delay time, τ for decay time (especially in the polymer literature), and subscripts without parentheses to designate first and second order autocorrelation functions such as g_1 and G_2.

There is a simple relationship (though difficult to derive[23]) between $g^{(1)}$, the normalized 1st-order (field) correlation function and $g^{(2)}$, the normalized 2nd-order (intensity) correlation function, namely the Siegert relationship:

$$g^{(2)}(\tau) = 1 + \left| g^{(1)}(\tau) \right|^2 \tag{III-3}$$

The relationship is only true if the scattered field is a Gaussian random variable, and this is true only if there are enough particles contributing to the total scattered light. In statistics, according to the Central Limit Theorem, even if individual values follow a different probability distribution, if there are something like 30 or more, the sum follows Gaussian statistics. The sum is a Gaussian random variable. Now the scattered electric field is a sum over scattering from individual particles and many (30 is about the minimum theoretically) contribute to the sum. Thus, DLS is never a single particle counter.

---

[23] Mandel, L, "Fluctuations in Light Beams", *Progress in Optics,* Vol. 2, Wolf, E., Ed., Wiley-Interscience, New York, 1963, p. 181.

III.4 What Is a Correlation Function?

Correlation is a measure of the similarity between two quantities. In the time domain it is useful, among other things, for establishing correlation between random signals and for establishing periodicity in the presence of noise. The correlation between two analog signals A(t) and B(t) is defined as:

$$C(\tau) = \lim_{T \to \infty} \left(\frac{1}{T}\right) \int_{t_0}^{t_0+T} A(t) \cdot B(t - \tau) dt \qquad \text{(III-4)}$$

T is the integration time; t_o is the initial time; A(t) and B(t) are the amplitudes of the time-varying signals; and C(τ) is called the cross-correlation function. A correlation function is a plot of C(τ) versus τ, where τ is called the delay time (also called the shift time or the lag time). If these signals arise from scattered light, then $C(\tau) = C_I(\tau)$. Here, and in the next few equations, when no subscript is given, reference is made to the general idea arising from any type of signal, not just scattered light intensity.

If A(t) and B(t) are the same function, that is B(t-τ) is simply a delayed version of A(t), then C(τ) is called the autocorrelation function, abbreviated ACF. The Fourier transforms of the autocorrelation and cross-correlation functions are the power spectral density and the cross power spectral density functions, respectively. As such, they contain, in principle, information in the frequency domain equivalent to the information the correlation functions contain in the time domain. See equations III-1 and III-2.

For stationary signals (signals independent of the initial time) and for ergodic systems (systems in which time averages are equivalent to ensemble averages), the correlation function can be written as:

$$C(\tau) = \langle A(0) \cdot B(t - \tau) \rangle = \lim_{T \to \infty} \frac{1}{T} \int_0^T A(t) \cdot B(t - \tau) \, dt \qquad \text{(III-5)}$$

where <A(0)·B(t-τ)> is the ensemble average.

The cross- or autocorrelation functions may be obtained with analog processing by multiplying the signal A(t) with a delayed version of signal B(t-τ) or itself, A(t-τ), and averaging this product over T seconds using, for example, an RC or operational amplifier integrator. However, analog integration can lead to distortion in the measured function. Because of the advances in digital electronics (1960's and onward), and since some important time-varying signals are discrete by nature (photons, for example), it is useful to approximate the integral by a finite sum of N products obtained by sampling the signal in discrete intervals. This discrete interval or sampling time (occasionally called the bin time) is designated as Δt.

In linearly spaced correlators, the delay time is an integer multiple of the sampling time. The last delay time equals the sampling time multiplied by the number of hardware correlator channels. Because of the proportionality between the delay and the sampling times, these terms were often, and erroneously, used interchangeably. Sometimes the term sample time was used, even though such use is confusing and incorrect.

In modern correlators, while linear-spacing is still possible, the delay time and the sampling time are de-coupled, allowing for a much greater range of delay times with fewer channels.

When approximating the ACF by a sum of N products, also called the number of samples, the ACF is written as:

$$C(\tau_j) = \lim_{N \to \infty} \frac{1}{N} \sum_{i=1}^{N} n_i \cdot n_{i-j} \qquad j = 1, 2, 3 \ldots M \qquad \text{(III-6)}$$

where τ_j is the jth delay time, n_i is the number of pulses during the sampling time Δt centered at some time t, n_{i-j} is the number of pulses during Δt centered at t-τ, and M is the number of correlator channels. Of course, when counting photons (pulses), the average over time, <n>/T, is in fact the average intensity of scattered light, <I>.

III.5 How is an Autocorrelation Function Accumulated?

See Figure III-1. It depicts a linear spacing of delay times τ using equal sampling (bin) times Δt. The delay time is an integer multiple of the sampling time.

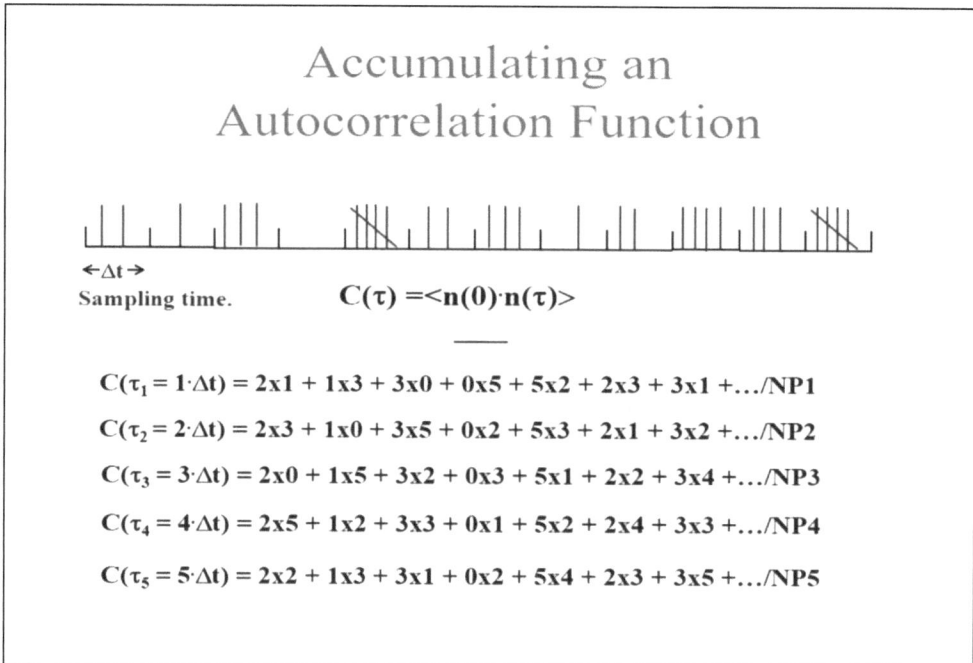

Figure III-1: Measurement (accumulation) of an autocorrelation function.

The signal is indicated by a fluctuation in the number of photo pulses registered in each bin: two in the first, one in the second, three in the third, zero in the fourth, five in the fifth, two in the sixth and so on. The ACF is the average of the product, $<n(0) \cdot n(\tau)>$, which is the sum over the product of the number of pulses at zero and a delay time later, divided by the number of products, NPX, where X is the channel number.

The first channel shows all products one delay time ($1 \cdot \Delta t$) apart divided by NP1. The second channel shows all products two delay times ($2 \cdot \Delta t$) apart divided by NP2. The third channel shows all products three delay times ($3 \cdot \Delta t$) apart divided by NP3, and so on.

Assuming the particles are rigid, globular, and all about the same size, it will be shown in the next section that the autocorrelation function (ACF) is a single exponential decay riding upon a substantial baseline. Using the scheme above, the graph is shown in Figure III-2.

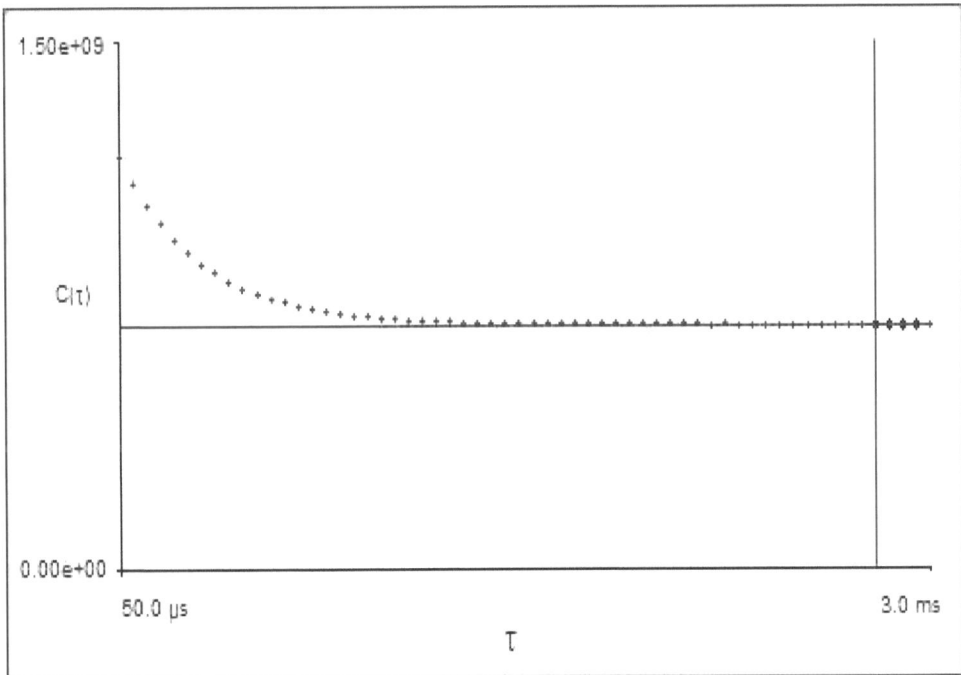

Figure III-2: An unnormalized ACF with linear channel spacing.

With modern correlators, and theory that shows linear spacing is unnecessary, most ACF's looks like the following (Figure III-3):

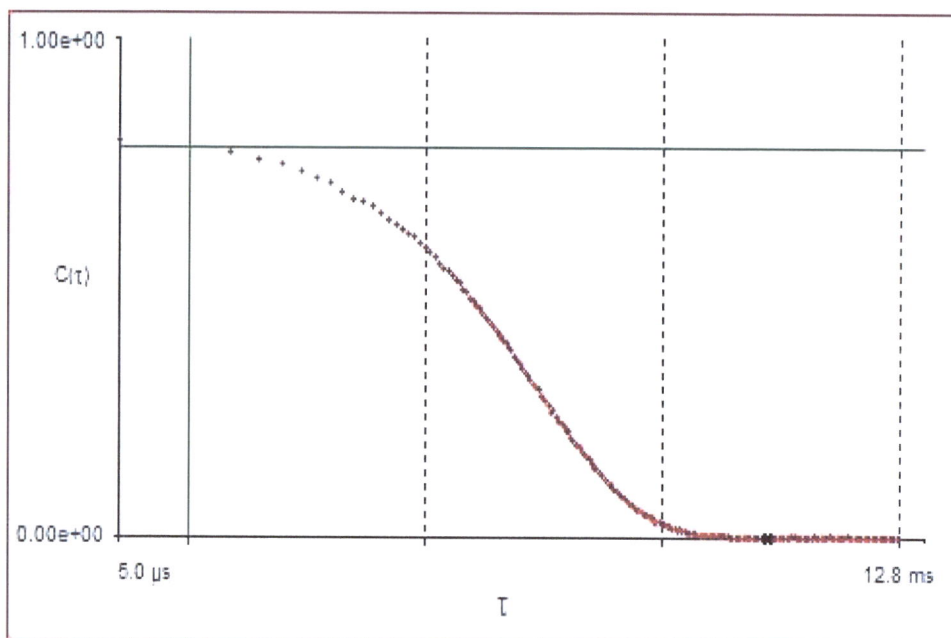

Figure III-3: Single exponential ACF with normalization and nonlinear channel spacing.

III.6 $g^{(1)}(\tau)$ And $P_E(\omega)$ For a Single Decay Process Arising from Diffusion

The full development[24] is more advanced than the presentation here allows. However, a few steps are worthy of presentation. The normalized, first-order autocorrelation function is related to a probability distribution function $G(\mathbf{R},\tau)$, where $\mathbf{R}$ refers to a three-dimensional displacement of a particle in time τ, by the following Fourier transform:

$$G(\mathbf{R},\tau) = \frac{1}{(2\pi)^3} \int g^{(1)}(\mathbf{q},\tau)e^{-i\mathbf{q}\cdot\mathbf{R}}d^3q \qquad \text{(III-7)}$$

This is a triple integral over all possible wave vectors $\mathbf{q}$ and the explicit dependence of the scattered electric field, and its correlation function are included along with τ. Usually, it is the time dependence that is determined via a measurement of the intensity autocorrelation function at a known $\mathbf{q}$ and so only the τ dependence is written. But for this development to proceed, both $\mathbf{q}$ and τ need to be shown.

Since $G(\mathbf{R},\tau)$ is a probability distribution function, $G(\mathbf{R},\tau)d^3R$ is the probability of finding a particle in the small region defined by d^3R centered at $\mathbf{R}$ and at time τ given that it was at the origin at $\tau = 0$. Now particles make slight changes in position as time involves. They do so randomly and according to the theory of random walk, $G(\mathbf{R},\tau)$ is the solution to Fick's

---

[24] Berne, B. J. & Pecora, R., "Dynamic Light Scattering with Applications to Chemistry, Biology & Physics, pages 56-60, Wiley-Interscience, 1976.

second law of diffusion:

$$\frac{\partial}{\partial \tau} G(\boldsymbol{R}, \tau) = D_o \nabla^2 G(\boldsymbol{R}, \tau) \tag{III-8}$$

D_o is the coefficient of self-diffusion in the limit of infinite dilution. In DLS, it is the mutual diffusion coefficient that is determined; however, at infinite dilution, it is equal to D_o.

The spatial Fourier transform of equation III-8 is:

$$\frac{\partial}{\partial \tau} g^{(1)}(\boldsymbol{q}, \tau) = -D_o q^2 g^{(1)}(\boldsymbol{q}, \tau) \tag{III-9}$$

Now this 1st-order, partial differential equation is easy to solve given the initial condition that $g^{(1)}(\boldsymbol{q}, 0) = 1$, because at zero delay time the particle hasn't moved, and the correlation function hasn't decayed. Its normalized value is unity. Therefore:

$$g^{(1)}(\boldsymbol{q}, \tau) = exp(-D_o q^2 \tau) \tag{III-10}$$

Typically, it is the translational diffusion coefficient D_o at infinite dilution that is desired, since it can be related to particle size. It is common to set $\Gamma = D_o \cdot q^2$, where Γ is called the linewidth. This reference to a spectral linewidth becomes more obvious when the electric field power spectral density function is calculated.

Using equation III-1, the normalized electric field power spectral density function can be calculated from equation III-10 as follows:

$$P_E(\omega) = \frac{1}{\pi} Re \int_0^\infty e^{-\Gamma \tau} e^{-i\omega\tau} d\tau = \frac{\Gamma/\pi}{\Gamma^2 + \omega^2} \tag{III-11}$$

This is a Lorentzian distribution and is typical for spectral line shapes. Its maximum occurs at $\omega = 0$ where $P_E(\omega) = 1/\pi\Gamma$. The width of the distribution is characterized by the half-width at half-maximum, $\Delta\omega_{1/2}$. Simple algebra shows that $\Delta\omega_{1/2} = \Gamma$, which is appropriately called the linewidth. The name was carried over when measurements of the intensity auto-correlation function replaced power spectral density measurements.

A Lorentzian function is shown in Figure III-4. Note that frequencies less than zero have no physical meaning. The arrow at half-maximum designates the half-width. The value of ω corresponding to the tip of the arrow is $\Delta\omega_{1/2} = \Gamma$.

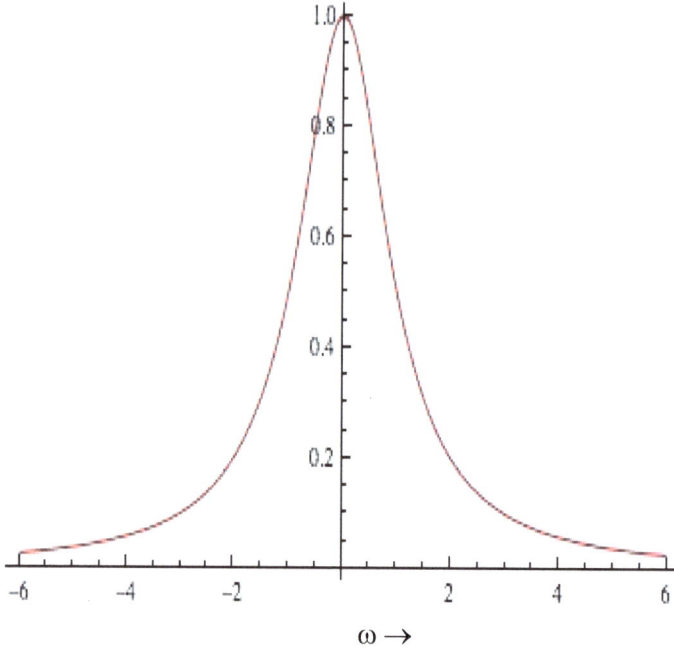

Figure III-4: A Lorentzian line shape with $\Gamma = 1$. Note: Only values for $\omega \geq 0$ are meaningful in light scattering.

III.7 $g^{(2)}(\tau)$ And $P_I(\omega)$ For a Single Decay Process Arising from Diffusion

From equations III-3 and III-10, we find:

$$g^{(2)}(\tau) = 1 + exp(-2D_o q^2 \tau) = 1 + exp(-2\Gamma\tau) \qquad \text{(III-12)}$$

Thus, the normalized, 2nd-order (intensity) ACF decays twice as fast as the 1st-order (electric field) ACF. The intensity power spectral density function, $P_I(\omega)$ is obtained from the Fourier transform and is given by:

$$P_I(\omega) = \delta\omega + \frac{2\Gamma/\pi}{(2\Gamma)^2 + \omega^2} \qquad \text{(III-13)}$$

The 1st term is a delta function at $\omega = 0$, a spike. The 2nd term is a Lorentzian with a half-width at half-height of 2Γ. There is a 3rd term, not shown, due to shot noise from the detector that represents a small, constant (dc) baseline.

So far, the baseline used for normalization of the measured ACF has been neglected as has the coherence properties. When included equation III-12 becomes:

$$C(\tau) = \lim_{T \to \infty} \left(\frac{1}{T}\right) \int_{t_0}^{t_0+T} A(t) \cdot B(t - \tau) dt \qquad \text{(III-14)}$$

Where B and f are discussed in more detail in the section below entitled "Introduction to Data Analysis".

III.8 Optical mixing (Homodyning) and the Meaning of Quasi-Elastic Light Scattering

The frequency of the light is on the order of 10^{+14} Hz. But no spectrum analyzer responds to such a high frequency. Consider a simple case of scattered electric fields from two particles, each from monochromatic light (thus, coherent) and colinear. Thus, when striking the square-law detector they add. The scattered intensity is proportional to the square of the total field:

$$I_s \propto |E_T|^2 = |E_1 \cos \omega_1 t + E_2 \cos \omega_2 t|^2 \qquad \text{(III-15)}$$

For simplicity, only the real part of the electric fields is considered and only the time dependence is made explicit. Expanding one finds:

$$I \propto |E_1|^2 \cos^2(\omega_1 t) + |E_2|^2 \cos^2(\omega_2 t) + 2|E_1 E_2| \cos \omega_1 t \cos \omega_2 t \qquad \text{(III-16)}$$

Now consider two trigonometric identities:

$$\cos^2 \alpha = \frac{1}{2}(1 + \cos 2\alpha) \text{ and } \cos \alpha \cos \beta = \frac{1}{2}\cos(\alpha - \beta) + \frac{1}{2}\cos(\alpha + \beta) \qquad \text{(III-17)}$$

Let $\alpha = \omega_1 t$ and $\beta = \omega_2 t$ and substitute equations III-17 into equation III-16. The result looks like this:

$$I(\omega) = \frac{1}{2}[|E_1|^2 + |E_2|^2] + |E_1 E_1| \cos(\omega_1 - \omega_2) \qquad \text{(III-18)}$$

$$+ terms \ in \ \cos(2\omega_1 t), \ \cos(2\omega_2 t), and \ \cos\big((\omega_1 + \omega_2)t\big)$$

The spectrum analyzer doesn't respond to terms that vary as ω_1, ω_2, and $\omega_1 + \omega_2$, because they all vary with too high a frequency. But the difference term $\omega_1 - \omega_2$ equals a beat frequency ω that can be followed. This is the essence of optical mixing spectroscopy as used in DLS. This is the value referred to in equations III-1, III-2, III-11, III-13 and Figure III-4, not the frequency of light.

Mixing can be accomplished in two ways. If only scattered light mixes on the detector, the technique is called the self-beating method (historically called homodyning). If a small amount of the laser light is redirected from the source and mixed with the scattered light in the right amount, the technique is called the local oscillator or reference beam method (historically called heterodyning). An example of the reference beam method is the basis of electrophoretic light scattering to determine zeta potential. It is explored in Chapter V.

What is the maximum width of a Lorentzian encountered in DLS? Since $\Gamma = D_o \cdot q^2$, the maximum corresponds to a minimum size and a maximum scattering angle. Ignoring the smaller effects that wavelength, liquid refractive index, viscosity and temperature have on the calculation, but using 1 nm for particle size and a high angle, $\Gamma_{max} \sim 5 \times 10^5$ s^{-1}. Thus, for the intensity power spectral density function the maximum width is $2\Gamma_{max} \sim 10^6$ s^{-1}: that is a MHz. The frequency of the light source is typically 10^{+14} Hz. The spreading in frequency of the monochromatic source due to diffusion is only 1 part in 10^{+8}. The frequency changes infinitesimally and this is what is meant by quasi-elastic, no measureable change in frequency and therefore energy during the collision between the photon and the particle.

III.9 Introduction to Data Analysis

The most common starting point in the analysis of the measured function, the second-order or intensity autocorrelation function (ACF) is the full Siegert relationship:

$$G^{(2)}(\tau) = B \cdot \left(1 + f^2 \left| g^{(1)}(\tau) \right|^2 \right) \qquad \text{(III-19)}$$

Here $G^{(2)}(\tau)$ is the un-normalized measured function, often designated as $C(\tau)$, where τ is the delay time. The baseline B is proportional to the square of the intensity. The amplitude f is related to the temporal and spatial coherence properties of the laser beam and the detector optics. It is roughly equal to the inverse of the number of detected coherence areas.[25] f is less than or equal to one and determines the intercept-to-baseline ratio. It is obtained during the fitting procedure and is here squared for convenience as it is always determined along with the normalized, first-order or electric field autocorrelation function $g^{(1)}(\tau)$. [Normalized correlation functions are traditionally shown as small g's and un-normalized ones as capital G's. Thus $f \cdot g^{(1)}(\tau) = \text{SqRt}\{G^{(2)}(\tau)/(B-1)\}$.] Physical parameters of interest are related to $g^{(1)}(\tau)$; whereas, the measured quantity is $G^{(2)}(\tau)$.

The Siegert relationship is rarely violated except when there are very few particles in the scattering volume and when the system is non-ergodic. The validity of the relationship rests on the assumption that the scattered electric field is a Gaussian random variable, and this is ensured when there are at least 30 particles at the intersection of the illuminated and detected volumes. [Note: Since the scattering volume is so small, 30 particles per scattering volume is typically 10^N/cm^3, where N is 7, 8, 9 or 10. In other words, this is not a single particle counting technique!] When this criterion is violated, number fluctuations occur and there is an additional, slowly decaying term in the correlation function. For more information on this arcane topic, consult one of the references. In Appendix DLS1, there is a brief discussion of

---

[25] Define A_{det} as the area of the detector (PMT or APD) determined by pinholes, lenses or single mode fiber optics. Define A_{coh} as the coherence area that can be estimated by λ^2/Ω where Ω is the solid angle subtended by the source at the detector. Then $N \approx A_{det}/A_{coh}$ and $f \approx A_{coh}/A_{det}$. Small pinholes decrease A_{det}, increase f as does focusing the beam so that A_{coh} increases. This maximizes the intercept-to-baselines ratio of the measured autocorrelation function.

concentration and how to estimate the number of particles that contribute to the scattered light.

A system is ergodic when the ensemble average of a property (theoretically predicted) equals the time average of the property (measured). Except for gels and colloidal glasses, or in the case of very strongly interacting systems, or particles trapped in a network, ergodicity is assumed for the cases of interest here. Consult the references for information on these non-ergodic systems and how, in some cases, static light scattering is combined with dynamic light scattering measurements to gain more information.

III.9.1 Monodisperse, Rigid, Globular Particles: Spheres

The references will show (see also equation III-10) that, in this case, a single exponential describes the correlation function:

$$\left| g^{(1)}(\tau) \right| = \exp(-\Gamma \cdot \tau) \tag{III-20}$$

Where, Γ is called the decay or relaxation rate and $\tau_r = 1/\Gamma$ is called the decay or relaxation time. For dilute systems where the decay is entirely due to translational, diffusive motion of the center of mass of the particle, it can be shown that:

$$\Gamma = D_o \cdot q^2 \tag{III-21}$$

Where, D_o is the translational diffusion coefficient and q is the magnitude of the scattering wave vector. This magnitude, discussed in Chapter II on SLS, and developed in Appendix SLS3, is given by:

$$q = \frac{4 \cdot \pi \cdot n_o}{\lambda_o} \cdot \sin(\theta/2) \tag{III-22}$$

Where, n_o is the refractive index of the liquid, λ_o is the wavelength of the laser in a vacuum, and θ is the scattering angle. A length scale is defined by q^{-1} against which distances between particles is compared. When the distance is large compared to q^{-1}, then self-diffusion is measured. When the distance is small, then mutual diffusion is measured. Mutual and self-diffusion vary differently with concentration (particle-particle separation) depending on the attractive and repulsive forces involved. However, for very dilute samples, both asymptotically approach the same limit, $D_T \rightarrow D_o$, where D_o is given by the Stokes-Einstein equation:

$$D_o = \frac{k_B \cdot T}{3 \cdot \pi \cdot \eta \cdot d_H} \tag{III-23}$$

Here k_B is Boltzmann's Constant, 1.3807×10^{-16} ergs/deg, T is the absolute temperature in Kelvin, η is the bulk viscosity of the liquid in which the particle moves, and d_H is the hydrodynamic diameter of the assumed sphere.

Hydrodynamic diameters include the particle "dry" diameter plus whatever is permanently attached to, and moves with, the surface. In the case of hard mechanical spheres with relatively short-chain surfactants added to impart stability (e.g. 100 nm PS and sodium dodecyl sulfate, approximately 2 nm), the difference between the "dry" and hydrodynamic diameters is usually not significant. However, when the particle is much smaller, and/or the attached species are much larger, as may be the case with long-chain polymer additives, then there may be a significant difference between the "dry" and hydrodynamic diameters. Interesting cases involve polyelectrolyte additives which exhibit isoelectric points. Such additives may change size significantly as a function of pH.

Sample Calculations

Calculate the diffusion coefficient in water at 25 °C for a 5-nm radius micelle.

$$D_T = \frac{k_B \cdot T}{3 \cdot \pi \cdot \eta \cdot d_H} = \frac{(1.381 \times 10^{-16} erg/degK) \cdot 298\ degK}{3 \cdot \pi \cdot (0.8904 \times 10^{-2} g/cm \cdot s) \cdot 10x\ 10^{-7} cm}$$

$D_T = 4.904 \cdot 10^{-7}$ cm^2/s

Calculate the scattering wave vector in backscatter at 173° using a 640-nm wavelength diode laser.

$$q = \frac{4 \cdot \pi \cdot n_o}{\lambda_o} \cdot \sin\left(\frac{\theta}{2}\right) = \frac{4 \cdot \pi \cdot 1.331}{640 \cdot 10^{-7} cm} \cdot \sin\left(\frac{173}{2}\right) = 2.609 \cdot 10^{+5} cm^{-1}$$

Calculate the decay rate.

$\Gamma = D_T \cdot q^2 = 4.904 \cdot 10^{-7} cm^2/s \cdot (2.609 \cdot 10^{+5} cm^{-1})^2 = 3.337 \cdot 10^{+4}$ s^{-1}

Calculate the relaxation time.

$\tau_r = 1/\Gamma = 29.97$ μs ≈ 30 μs

III.9.2 Self-Beating, Local-Oscillators, Reference Beam and Homo/Heterodyning

In equation III-14 it was assumed that only the scattered light mixed on the detector to produce self-beating. This gives rise to the square term in $G^{(2)}$, and, in the case of a monodisperse system of rigid particles, the resulting $\exp(-2\Gamma\cdot\tau)$.

Imagine instead that a small fraction of the incident beam intensity (routed around the sample cell) was also mixed with the scattered light on the detector. This small fraction is called a local oscillator and sometimes a reference beam. It does not have the resulting time dependence of the scattered light. Provided that the strength of the local oscillator is greater than a few times the scattered light, but not so great as to swamp the scattered light completely, $G^{(2)}$ varies in proportion to $g^{(1)}$ rather than its square. (In this case, the data analysis should proceed by fitting without taking the square root.) This is called the local oscillator or reference beam mode and is rarely used in DLS. It is used in the case of electrophoretic and phase analysis light scattering where the sign of the electrophoretic velocity is required. From it the sign of electrophoretic mobility and zeta potential are determined. This is explored further in Chapter V.

However, unintentional stray light can also act as a local oscillator. Also called flare light, it occurs at extreme angles where the light striking the entrance and exit of the cell encounters the biggest differences in refractive index: air/cell-wall interface. Surrounding the cell with a liquid whose refractive index closely matches that of the cell, moves the air/surrounding vat-wall interface much further away from the detection system. (The vat is a much larger diameter glass or quartz container that holds the index matching liquid.) The method is called index matching. However, there is still the sample-liquid/cell-wall interface. Light from this interface which can only be reduced with a larger diameter cell or by tilting the cell such that the back-reflected and flare light is minimized. When small volumes are required, tilting is the preferred method.

In much of the DLS literature, and in the general references A and B cited at the beginning of this section, it is common to refer to the self-beat mode as homodyning and the local oscillator mode as heterodyning. In reference D, it is pointed out that homodyning should refer to a local oscillator obtained from the same light source whereas heterodyning should refer to a local oscillator obtained from a different light source. Thus, most of the literature incorrectly refers to self-beating as homodyning and homodyning as heterodyning. These unfortunate misnomers are wide spread.

If the data is collected such that the self-beat mode dominates, but analyzing the data using the local oscillator (reference beam) will result in sizes *one half the expected values* and vice-versa. Thus, it is important that the data analysis matches the design of the instrument.

In a more detailed analysis of the local oscillator mode one finds that the restriction of having many particles is lifted. Thus, number fluctuations are no longer a problem in the local oscillator mode with a small number of particles. However, given the difficulty of setting a local oscillator correctly and the naturally low intercept-to-baseline ratio that results, the local oscillator mode is rarely used in practice except as noted above for electrophoretic and other velocity measurements. [The intercept-to-baseline ratio is defined as $C(\tau = 0)/B$, where, since $C(0)$ is not measured, it is estimated by extrapolating the first few channel contents to $\tau = 0$.]

In everything that follows it is assumed that the self-beat mode is used.

III.10 Model Fits

These include cumulants, Williams-Watts, and the double exponential.

III.10.1 Cumulants

The best-known method for analyzing polydispersity is the method of cumulants.[26] For an assumed continuous, normalized distribution of decay rates $G(\Gamma)$:

$$\left| g^{(1)}(\tau) \right| = \int_0^\infty G_I(\Gamma) \, exp(-\Gamma \cdot \tau) \, d\Gamma \qquad \text{(III-24)}$$

$$\int_0^\infty G_I(\Gamma) \, d\Gamma = 1 \qquad \text{(III-25)}$$

Where normalization is ensured by equation III-25.

Do not confuse $G_I(\Gamma)$ with either $G^{(1)}(\tau)$ or $G^{(2)}(\tau)$. The first is the intensity-weighted distribution function of Γ's from which the distribution of diffusion coefficients, size or molecular weights is ultimately determined. The second and third represent the unnormalized electric field and intensity autocorrelation functions, respectively:

$$G_I(\Gamma) = \frac{\sum_i I_i \cdot \delta(\Gamma - \Gamma_i)}{\sum_i I_i} = \frac{\sum_i N_i m_i^2 P_i \cdot \delta(\Gamma - \Gamma_i)}{\sum_i N_i m_i^2 P_i} = \frac{\sum_i c_i M_i P_i \cdot \delta(\Gamma - \Gamma_i)}{\sum_i c_i M_i P_i} \qquad \text{(III-26)}$$

For a discrete distribution of particles or molecules, the distribution is written as, I_i is the intensity scattered by, m_i is the mass of, and N_i is the number per scattering volume of the ith particle with decay rate Γ_i. Also, if the particles are molecules, readers will recognize that c_i is the mass concentration of a polymer molecule with molecular weight M_i. The term $\delta(\Gamma - \Gamma_i)$ is the delta function and is used mathematically to select discrete values when integrated in equations III-24 and III-25.

---

[26] First proposed by D.E. Koppel, *J. Chem. Phys.*, **57**, 4814, 1972.

The factor P_i accounts for angular scattering effects for particles larger than about $\lambda/20$, the usual approximation for a non-Rayleigh particle. P_i is calculated from Mie scattering theory and requires the particles' refractive index. For Rayleigh particles and at sufficiently low angles $P_i = 1$, and experienced readers will realize that the formalism above will lead to a z-average diffusion coefficient from which an inverse, z-average particle diameter is calculated. More generally the averaging is by intensity and the resulting particle size is an inverse, intensity-weighted diameter. (It is sometimes claimed, incorrectly, that DLS measurements *always* yields a z-average result. This is only true when $P_i = 1$. See Appendix DLS2.)

Expanding the exponential in equation III-24 about an average value, and integrating term-by-term one obtains:

$$\left|g^{(1)}(\tau)\right| = \exp\left(-\bar{\Gamma}\tau\right) \bullet \left[1 + \frac{\mu_2}{2!}\cdot\tau^2 - \frac{\mu_3}{3!}\cdot\tau^3 + \frac{\mu_4}{4!}\cdot\tau^4 + ...\right] \qquad \text{(III-27)}$$

Here the first moment about the origin, the "average", is defined as:

$$\bar{\Gamma} = \int_0^\infty \Gamma \cdot G_I(\Gamma) \cdot d\Gamma \qquad \text{(III-28)}$$

The nth moment about this average value is defined as:

$$\mu_n = \int_0^\infty (\Gamma - \bar{\Gamma})^n \cdot G_I(\Gamma) \cdot d\Gamma \qquad \text{(III-29)}$$

Taking the natural logarithm of Eq. (III-27) and expanding in a Taylor series the term in brackets, under the assumption that the sum of the various terms involving powers of τ is small, one gets the following simple polynomial (after adding back the amplitude f):

$$\ln\left[f\left|g^{(1)}(\tau)\right|\right] = \ln(f) - \bar{\Gamma}\tau + \left(\mu_2/2!\right)\cdot\tau^2 - \left(\mu_3/3!\right)\cdot\tau^3 + \left((\mu_4 - 3\mu_2^2)/4!\right)\cdot\tau^4 + ...$$

$$\text{(III-30)}$$

The coefficient of τ^n is called a cumulant, though the first three are equal or proportional to simple moments.

The first cumulant, $\bar{\Gamma}$, is the most important term that is recovered. In polymer solutions, it can be shown that:[27]

$$\bar{\Gamma} = D_T \cdot R_g^{-2} \cdot \left(q \cdot R_g\right)^n \qquad \text{(III-31)}$$

---

[27] Doi and Edwards, The Theory of Polymer Dynamics, Clarendon Press, Oxford, U.K., 1986, page 107

Where, depending, on the magnitude of $q \cdot R_g$, n is either 2 or 3. R_g is the radius of gyration, a measure of polymer size. It is proportional to $R_H = d_H/2$, the hydrodynamic radius, but not equal to it. When n = 2, simple center-of-mass diffusive motion occurs, and $\overline{\Gamma} = \overline{D}_T \cdot q^2$. This occurs in dilute solution when $q \cdot R_g < 1$. When $q \cdot R_g > 1$, n = 3; the motion observed in DLS is not diffusive; the internal polymer modes are probed; and $\overline{\Gamma}$ varies as q^3. Furthermore, $\overline{\Gamma}$ is then no longer simply dependent on the particle size or molecular weight.

For slightly concentrated systems the measured and infinitely dilute diffusion coefficients are expressed as:

$$\overline{D}_T = \overline{D}_o \cdot (1 + k \cdot c + \ldots) \tag{III-32}$$

This equation can also be applied to suspensions of particles as well. However, in more concentrated suspensions of particles or solutions of polymers, the q^2 modes observed are not proportional to self-diffusion but to mutual (also called cooperative) diffusion. Self- and mutual diffusion vary differently with concentration, but this is a topic beyond the scope of this presentation. Suffice it to say that it is wrong to interpret every ACF in terms of particle size derived from a self-diffusion coefficient.

Assuming the ACF can be interpreted in terms of self-diffusion, then for particle sizing the intensity-weighted, self-diffusion coefficient is an average. After invoking the Stokes'-Einstein equation III-23, the following applies:

$$Effective\ Diameter \equiv \left(\overline{\frac{1}{d_H}}\right)^{-1} = \frac{\sum_i N_i d_i^6 P_i}{\sum_i N_i d_i^5 P_i} \tag{III-33}$$

The **Effective Diameter**, abbreviated Eff. Dia., is neither the mass nor volume nor number-weighted diameter. Nor is it the intensity-weighted diameter. Nor is it, as it is sometimes erroneously called, the z-average diameter. (Again, see Appendix DLS2.) The intensity weighting is on the diffusion coefficient. Since that varies inversely with particle size, equation III-33 results. (Note: The d^6 term arises from the m^2 term in Eq. (III-26) but is no surprise after reading Chapter II on SLS.)

The Eff. Dia. is normally larger than most other common, average diameters, since the averaging is over the 6th and 5th powers. In certain cases, were P_i is a strong function of the angle, and a minimum in the Mie scattering pattern obtains, it is possible to observe a reversal such that the Eff. Dia. is smaller than the volume average diameter. But generally, it is the other way around. This explains why dust and other, unwanted large particles, while few, can distort the DLS results.

The second moment μ_2 is the intensity-weighted variance of the diffusion coefficient distribution when samples are dilute and interactions negligible. The relative variance, $\mu_2/\overline{\Gamma}^2$, also called the polydispersity index (PDI) or Poly (short for polydispersity), is a measure of the width of the intensity-weighted, decay-rate distribution, $G_1(\Gamma)$. [In early literature references this relative value is called the reduced second moment.] Therefore, when the distribution is one of particles, Poly is a measure of the size distribution width. It is the intensity-weighted relative variance of the diffusion coefficient, but nonetheless a measure of width. When the Poly equals zero, the sample is monodispersed, and as Poly increases so does the width of the distribution. It applies to monomodal as well as multimodal distributions since moments do not indicate distribution shape, only magnitudes. Sometimes that is enough since a sample such as a pure monoclonal antibody, a latex used as a reference material, or any other sample that is supposed to be narrowly distributed must have a low Poly, typically ≤ 0.025.

The validity of the cumulant expansion rests on the assumption that the higher order terms do not contribute significantly. As a generalization, it is found that the cumulant analysis works best when Poly ≤ 0.3, approximately.

A relative third moment, defined as the Skew $= \mu_3/(\mu_2)^{3/2}$, is a dimensionless number that can, if the skew is repeatable, indicate if the distribution is skewed to the left (negative skew) or to the right (positive skew) of the modal value.

A relative fourth moment, defined as the Kurtosis $= \mu_4/(\mu_2)^2$, is a dimensionless number that, for unimodal distributions only, if repeatable, indicates the relative peaked-ness of the distribution compared to a Gaussian distribution (Kurtosis $= 3$). The 4th moment obtained using the cumulant method is rarely repeatable and therefore of little value. It is included here for historic purposes only.

III.10.1.1 Using the Cumulant Fit: Interpretation

See Table III-2 for an example of a sample correlation function successively fit to 1st, 2nd, 3rd, and 4th order.

	Gamma (s^{-1})	Eff. Diam. (nm)	Poly	Skew	Kurtosis	RMS Error	Amplitude
Linear:	1.477e+03	116.1				9.3404e-04	0.748
Quadratic:	1.500e+03	114.3	0.039			1.5591e-05	0.750
Cubic:	1.500e+03	114.3	0.041	0.32		3.5361e-06	0.750
Quartic:	1.500e+03	114.3	0.040	0.06	1.48	4.7423e-05	0.750

Table III-2: Successive cumulant fits to the same ACF.

A standard polynomial least squares fitting procedure is used with weighting since the fit is to the logarithm of $g^{(1)}$ rather than the directly measured value $G^{(2)}$.

The various levels of fit are listed in the first column.

The linear fit just includes the linear term in τ. From the fitted Γ (capital "Gamma"), the diffusion coefficient can be calculated from equations III-21 and III-22. The Eff. Dia. is calculated from equation III-23. To determine **Poly** requires at least a second order fit; to determine the **Skew** requires at least a third order fit; and to determine the **Kurtosis** requires at least a fourth order fit.

The RMS error is the root-mean-square error. It is the square root of the sum of squares of the differences between the fitted ACF and the measured. Before taking the square root, the sum is divided by the number of data points minus the number of fitted parameters. Smaller RMS errors means a better fit in the case of cumulants.

The amplitude is f^2 as in equation III-19. It represents the amplitude above the baseline chosen for the fit. The amplitude is a number between 0 and 1. Here the value of 0.75 means the detected and coherence areas were reasonably well matched.

A second order or quadratic fit yields the **Poly** from which distribution width can be decided. The results above show a **Poly** of 0.039 and 0.041 for 2nd and 3rd order fits, respectively. The RMS values are lower than a 1st order fit, meaning the fit was better. No doubt the sample is not monodisperse in this case.

The 4th order fit, judging by a significant increase in RMS, is worse than the 3rd order fit, which in turn is better than the 2nd order fit. This suggests that no more than three distribution parameters (ignoring f) can be usefully determined. For example, if another algorithm (CONTIN, NNLS, Exp. Sampling, for example) yielded a bimodal (two peak positions plus the ratio of amount in each peak), this is consistent with three parameters. A trimodal is not consistent (three peak positions and ratio of two independent areas, for a total of five distribution parameters). A broad unimodal distribution is also consistent with three distribution parameters.

Higher order fits are not always worthwhile and may lead to large RMS errors if the data cannot support the forced fit. Look for trends in the fitted parameters vs. fit order that are repeatable and consistent with any prior knowledge of the scattering system.

Do not expect the cumulant method to work well when the distribution is very broad. This applies to broad particle size and polymer distributions, and it also applies to scattering from complex fluids and polymer solutions where more than diffusion is occurring.

III.10.1.2 Lognormal Approximation from Cumulants

While the method of cumulants is the oldest and still a quite useful technique for analyzing ACFs, it cannot yield the shape of the decay rate distribution; therefore, it cannot shed light on the shape of the size distribution. It can only yield, at best, moments of the distribution. A simple approximation, however, allows one to convert the first two moments into an intensity-weighted size distribution which can then be plotted. For particle size distributions, a common unimodal choice is the lognormal curve, which is a Gaussian or Normal curve in log space.

The nth moment of an x-weighted, lognormal distribution of particle diameter is given by:

$$\overline{d_x^n} = d_{gx}^n \cdot \exp\left(\frac{n^2}{2} \cdot \ln^2 \sigma_g\right) \qquad \text{(III-34)}$$

Where x signifies the weighting (intensity, volume, surface area, number), d_{gx} is the geometric mean diameter with x weighting (also equal to the median value on a log scale), and σ_g is the geometric standard deviation (also known as the harmonic standard deviation). A unique feature of a lognormal distribution is that its width, characterized by the parameter σ_g, is independent of the weighting.

The entire lognormal curve by intensity is determined with both d_{gI} and σ_g specified. Set the Eff. Dia. = d_{gI}; then calculate σ_g using the simplifying assumption that $P_i = 1$. The calculation is outlined as follows.

The Poly, Q, can be written as: $\qquad Q = \left(\overline{D_T^2}/\overline{D_T}^2\right) - 1 \qquad \text{(III-35)}$

where the subscript "I" means intensity weighting. Using the definition in Eq. (III-35) and since the intensity "I" is proportional to $N \cdot d^6 \cdot P$, where P is the Mie scattering factor which reflects any angular dependence, one can relate Q to the intensity-weighted moments and thus to the Lognormal distribution. Further progress requires values for P. Though strictly true only for Rayleigh particles or at low angles, the simplifying assumption that $P = 1$, allows one to derive the following simple formula relating the measured polydispersity index Q and the geometric standard deviation, σ_g:

$$\sigma_g = \exp\{[\ln(1 + Q)]^{1/2}\} \qquad \text{(III-36)}$$

Please note that the Lognormal fit from cumulants cannot possibly describe any multimodal distribution, nor one that is skewed on a log scale, nor was it intended as anything more than a way to visualize graphically the relative width as contained in Poly. With the two parameters that define the Lognormal equation, any number of other values can be calculated. Yet, there is no more information than that contained in the two values: Eff. Dia. and Poly. The Lognormal distribution, and its assumptions stated above, may have value as a quality control tool only, nothing more.

III.10.2 Williams-Watts (W-W)

This function, also known as a stretched exponential, and related to the Tung-Weibull distribution in statistics, is useful when the ACF is a broad, non-exponential function such as those found when measuring complex fluids like viscous liquids and gels.[28,]

---

[28] For more information consult Section 2, <u>Photon Correlation of Bulk Polymers</u>, page 127, G.D. Patterson, **Light Scattering from Polymers,** Springer-Verlag, 1983.

The Williams-Watts function is given by:

$$\left| g^{(1)}(\tau) \right| = \exp\left[-\left(\frac{\tau}{\tau_r} \right)^\beta \right]$$

(III-37)

With $0 < \beta \leq 1$ describing the width of the distribution of relaxation times τ, and τ_r is a characteristic relaxation time. When $\beta = 1$, the distribution is exponential, but as β approaches zero the decay in the ACF covers many decades.

An average decay rate can be calculated by integrating equation III-37 multiplied by τ. The result is:

$$\bar{\tau} = \frac{\tau_r}{\beta} \cdot \Gamma\left(\frac{1}{\beta} \right)$$

(III-38)

Here $\Gamma(1/\beta)$ represents the gamma function tabulated in statistical textbooks, not the linewidth associated with the decay of an ACF.

The Williams-Watts function is interesting in that it is used to fit the $g^{(1)}$ function directly rather than a size distribution.

III.10.2.1 Using the Williams-Watts Fit: Interpretation

An example of a Williams-Watt fit is shown below Table III-3:

Amplitude:	0.944
Linewidth, $\Gamma = 1/\tau_R$:	5.147e+02
Relaxation Time, τ_R:	1.943e-03
Width Parameter, β:	5.335e-01
RMS Error:	1.774e-02

Table III-3: Example of William-Watts fit.

The W-W function is fit to the data using a non-linear least squares approach, a standard Marquardt-Levenberg algorithm. No weighting is required. If the problem had been linearized before fitting, by taking the log of the log, weighting would be required. The log of the log is notorious for removing much of the interesting variations in a function and, sometimes, leads to results devoid of much physical meaning. In any case one must approach the W-W fit as a convenient numerical fit to the data as fundamental theory showing the appropriateness of the W-W equation is sparse.

III.10.3 Double-Exponential (D.E.)

Like the W-W and cumulant fits, the D.E. fit assumes a specific form for the first-order ACF, namely:

$$|g^{(1)}(\tau)| = G_I(\Gamma_1) \cdot exp(-\Gamma_1\tau) \quad + \quad G_I(\Gamma_2) \cdot exp(-\Gamma_2\tau) \qquad \text{(III-39)}$$

Where $G_I(\Gamma_i)$ represents the relative intensity contributed by the ith particle with decay rate Γ_i. This too is a non-linear least squares problem for which the Marquardt-Levenberg algorithm is used.

Note that this is a forced fit: do not be surprised when a bimodal answer is obtained. One can fit reasonably well some very broad but unimodal distributions using a double exponential. Nothing can be said about the actual distribution shape when force-fit to a double exponential. The global mean and global variance, as determined by both resulting peaks, are useful for characterizing moments of, but not for determination of, the true distribution shape. One could not say, for example, because the RMS error is small (meaning a good fit) that the distribution is, therefore, a double exponential.

III.10.3.1 Using the Double Exponential Fit: Interpretation

An example of a D. E. is shown in Figure III-5 below in Γ-space:

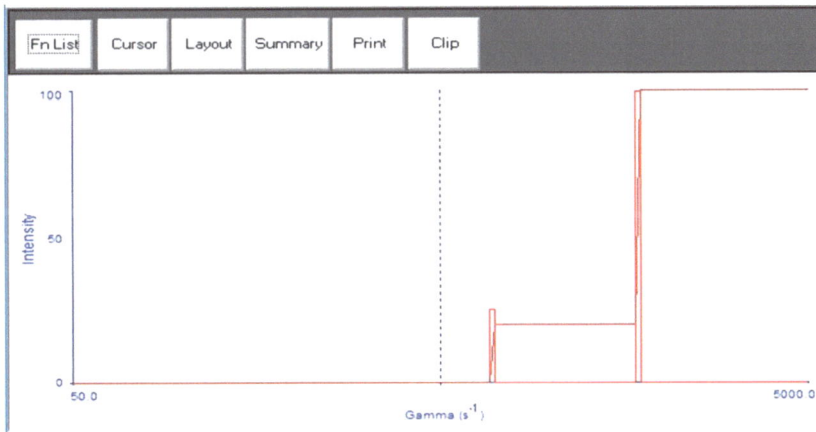

Figure III-5: Double exponential fit in Γ-space from a measured ACF.

The vertical bars are centered at Γ_1 and Γ_2. The thickness of the bar has no meaning. It was chosen only for display purposes. The relative heights of the bars represent $G_I(\Gamma_1)$ and $G_I(\Gamma_2)$. The single line starting at zero, connecting the bars, and eventually rising to 100 represents the cumulative distribution of the relative intensity-weighted Γ distribution. It is a step function here. Alternatively, the results can be transformed to diameter-space as shown in Figure III-6.

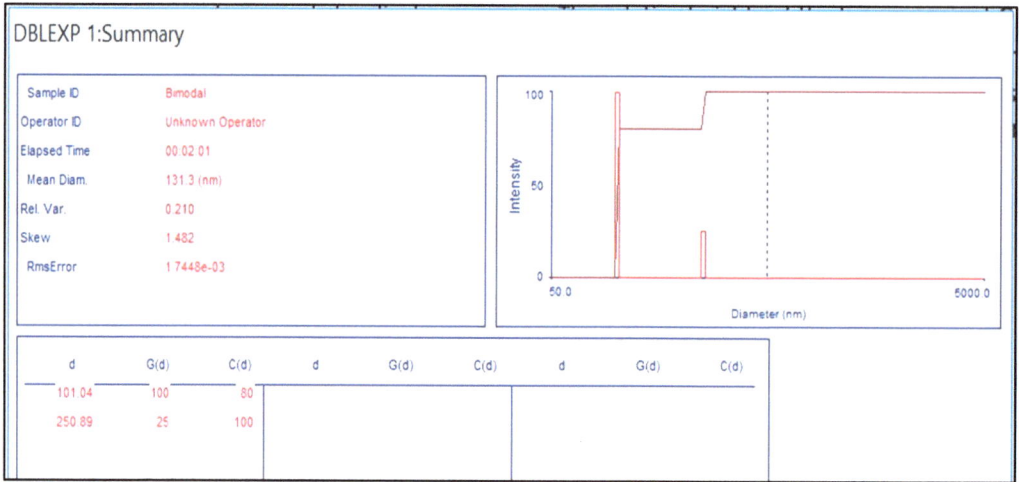

Figure III-6: A D.E. fit always results in a bimodal if it converges. Here the intensity-weighted tabular and graphical results are shown.

The kth value of the cumulative undersize function, $C_y(d)$, is calculated from the histogram values $G_y(d)$ as follows:

$$C_y(x_k) = \frac{\sum_{i=1}^{k} G(x_i)}{\sum_{i=1}^{N} G(x_i)} \tag{III-40}$$

where the values of $C_y(x)$ are rounded to the nearest whole percent and N is the total number of entries. For a D.E., i = 1, 2 and N = 2. The variable x can represent either the decay rate Γ, or its inverse the relaxation time τ_r, or a diameter d calculated from Γ and Eqs. (III-21), (III-22), and (III-23). And the weighting is denoted by the subscript y that is either I (intensity), v or m (volume or mass), n (number), and rarely s (surface area). The other statistical parameters are calculated as follows:

$$Mean\ of\ x_y,\ \bar{x} = \frac{\sum_{i=1}^{N} x_i \cdot G_y(x_i)}{\sum_{i=1}^{N} G_y(x_i)} \tag{III-41}$$

$$Variance\ of\ x_y,\ \sigma^2 = \frac{\sum_{i=1}^{N}(x_i-\bar{x})^2 \cdot G_y(x_i)}{\sum_{i=1}^{N} G_y(x_i)} \tag{III-42}$$

$$Third\ moment\ of\ x_y,\ \mu_3 = \frac{\sum_{i=1}^{N}(x_i-\bar{x})^3 \cdot G_y(x_i)}{\sum_{i=1}^{N} G_y(x_i)} \tag{III-43}$$

Here, the third moment like the first, $\bar{x}$, , and the second squared, σ^2, is that of the y-weighted distribution, either intensity, volume or mass, number, and rarely surface area.

The Rel. Var. $\equiv \sigma^2/\overline{(x)}^2$ and the Skew $\equiv \mu_3/\sigma^3$ values are shown in Figure III-6.

Please note that the double exponential fit does not always converge. It depends on the noise and shape of the ACF as well as the initial guesses. In this case, for the initial guesses, the ACF is first fit to a second order cumulant producing $\bar{\Gamma}$ and μ_2. Then the initial guesses are as follows: $G_1(\Gamma_1) = G_1(\Gamma_2) = 0.5$, $\Gamma_1 = \bar{\Gamma} + \sqrt{\mu_2}$, $\Gamma_2 = \bar{\Gamma} - \sqrt{\mu_2}$. If any information is available for any of the four parameters (two G and two Γ values), that is *a-priori* information, and using it can improve convergence and fit.

III.10.3.2 Double Exponential Presentation: Relaxation Time, Particle Size

Starting with equation III-39, the presentation above covered the intensity weighting of the decay rate. But equation III-39 could have been written in terms of relaxation times $\tau_r = 1/\Gamma$. The results are obtained from the Γ-fit as follows: $\tau_{r1} = 1/\Gamma_1$, $\tau_{r2} = 1/\Gamma_2$, $G_1(\tau_{r1}) = G_1(\Gamma_1)$, and $G_1(\tau_{r2}) = G_1(\Gamma_2)$. The mean, relative variance and skew are then calculated from equations III-41, III-42, and III-43. The cumulative distribution is calculated from equation III-40.

The diameters, weighted by intensity, are obtained from the Γ-fit as follows: $d_1 = c/\Gamma_1$, $d_2 = c/\Gamma_2$, $G_1(d_1) = G_1(\Gamma_1)$, and $G_1(d_2) = G_1(\Gamma_2)$. The mean, relative variance and skew are then calculated from equations III-41, III-42, and III-43. The cumulative distribution is calculated from equation III-40. Here the constant c is $(k_BT/3\pi\eta)\cdot q^2$, where q is defined in equation III-22.

The diameters, weighted by volume, which is equal to weighted by mass if all the particles have the same density, are obtained from the intensity-weighted values $G_1(d)$. For historical reasons, a mass-weighted distribution is often referred to as the weight distribution. The results are obtained from the intensity-weighted distribution $G_1(d)$ by forming the unnormalized, volume-weighted value $G'_v(d) = G_1(d)/(d^3P(\theta))$, where $P(\theta)$, is the angular part of the Mie scattering coefficient.[29] A normalized version, G_v, is obtained by setting the highest value of G'_v equal to 100 and scaling the remainder proportionally.

Given the set of $G_v(d)$, the mean, relative variance, and skew are then calculated from Eqs. (III-23), (III-24), and (III-25). The cumulative undersize distribution is calculated from Eq. (III-22).

Finally, the diameters, weighted by surface area or weighted by number, $G_s(d)$ or $G_N(d)$, are related to the volume-weighted distribution as follows: $G'_s(d) = G_v/d$ and $G'_N(d) = G_v/d^3$, where the primes signify unnormalized values. Then normalized versions of G_s and G_N are calculated from the unnormalized versions. Again, once the sets of $G_s(d)$ and $G_N(d)$ are

---

[29] Mie coefficients are often calculated using a modified version of the program BHMIE for homogeneous spheres found in the appendix of the book **Absorption and Scattering of Light by Small Particles**, authors C.F. Bohren and D.R. Huffman, Wiley-Interscience publishers, 1983.

known, the respective means, relative variances, and skews are then calculated from equations III-41, III-42, and III-43. The cumulative undersize distributions are calculated from equation III-40. As discussed in Chapter I, such calculations for surface weighting assume the particles are not porous. This weighting is rarely used in DLS results.

It bears repeating: Just because the RMS and residual plots —channel-by-channel comparisons of the fitted and measured ACFs— indicate good agreement does not mean that the distribution is bimodal. Historically, the difficulty in getting good D.E. fits was the first indication that fitting sums of exponentials from DLS data was much more involved than first expected.

III.11 Fundamental Limit on Distribution Information from DLS

One of the first attempts to fit more than a monodisperse sample was mixing two monodisperse samples and using the D.E. fit. Since the two sizes were known, the two values Γ_1 and Γ_2 could be calculated. And, typically, the intensities were measured separately; thus, the two relative intensities $G_1(\Gamma_1)$ and $G_1(\Gamma_2)$ could also be calculated. In non-linear least squares fitting, initial estimates are used with an iterative procedure to find the final answers.
Not surprisingly, when the initial estimates were close to the true values, the fit quickly converged. But for estimates that diverged a little from the answers already known, the fits didn't converge or converged on the wrong answers. This was worse when the noise on the correlation function was larger and the baseline not well determined. It was worse when the two sizes differed less than about 2:1.

In simple linear least squares fitting, such as with a single exponential (monodisperse), with a polynomial in τ (cumulants), with the William-Watts, where the first two are linearized by fitting the log, and the third by fitting the log of the log, the fit proceeds by minimizing the sum S of the squares of the residuals (the basis for least squares fitting):

$$S = \Sigma_i w_i \cdot \left(g^{(1)}(\tau_i) - \Sigma_j G_l\left(\Gamma_j\right) \cdot exp\left(-\Gamma_j \cdot \tau_i\right) \right)^2 \qquad \text{(III-44)}$$

Where the summation over i represents the sum over the correlation channels at τ_i, and the summation over j represents that over the assumed number of decay rates Γ_j. This is a weighted sum, with w_i the weighting factor. It is related to the inverse of the variance at each channel.

In simple linear or linearized cases, there is a single, well-defined minimum in S versus the fit parameters. But for non-linear functions such as the D.E., this may not be the case. There may be local minima with only a single absolute minimum. Thus, if the initial estimates are too far off, the iteration gets trapped in a local minimum with the wrong results.

This was but a prelude when attempts were made to deconvolute the more general equation relating the measured $g^{(1)}(\tau)$ to $G_1(\Gamma)$. To be sure, it is $G^{(2)}(\tau)$ [that is $C(\tau)$] that is measured, but if the baseline is well-determined, $g^{(1)}(\tau) = SqRt\{G^{(2)}(\tau)/(B\text{-}1)\}$, aside from the optical constant f (so-called because of its relationship to the ratio of the detected to the coherence area). The more general problem is then to solve for $G_I(\Gamma)$. From it, the distribution weighted by intensity for Γ is determined. And once determined the intensity-weighted distributions for τ_r, D_T, d_H, and even M follow. (Calculating M in the case where the Mark-Houwink-Sakurada equation holds, though using GPC/SEC is a better technique for finding the molecular weight distribution.)

Thus, the general problem is reduced to a deconvolution of:

$$|g^{(1)}(\tau)| = \int_0^\infty G_I(\Gamma)\, exp(-\Gamma \cdot \tau)\, d\Gamma \qquad \text{(III-45)}$$

$$\int_0^\infty G_I(\Gamma)\, d\Gamma = 1 \qquad \text{(III-46)}$$

Those familiar with transforms will recognize the first equation as a Laplace transform with the exponential term as the kernel. Calculating $G_I(\Gamma)$ from the measured $g^{(1)}(\tau)$ is the equivalent of Laplace transform inversion.

Tables of inverse Laplace transform exist for many, many different kernels. But not for an exponential. Thus, it is called an ill-conditioned or ill-posed Laplace transform. There are other ill-conditioned Laplace transforms but this one, with an exponential as the kernel, is a prototype for the difficulties in dealing with them. [Note: Sums of exponentials also occur in circular dichroism measurements and spectra from mixtures of dye molecules with forbidden singlet-triplet transitions.] What are the consequences of this ill-conditioning and how does it limit size (and other) distribution information?

First there is no analytic solution, no simple functional form for the distribution. Thus, only numerical solutions are possible. If this were the only problem, living with a table of $G_I(\Gamma)$ vs. Γ would be quite acceptable. However, there is more to the limitation from inverting this ill-conditioned Laplace transform: minor differences in the noise along the correlation function, from one measurement to the next, can give rise to different results. Where one measurement might yield a unimodal set of values when plotted, the next might yield a bimodal or multi-modal. A slight difference in the measured baseline B used to calculate $g^{(1)}(\tau)$ can have the same results. And, finally, finding the minimum of S, the sum of the squares of the residuals, is no longer a guarantee of finding the right solution. Several solutions, all with approximately the same S, may exist but with different distributions.

Worse, it was also found that there was limited information to be found in such deconvolutions. For example, for a bimodal, each mode with a width, there are at least five parameters (degrees of freedom) necessary to describe it: two peak positions, two widths (a standard deviation or half-width at half height, both are widths), and the ratio of the area under each curve describing the relative scattered intensity for each peak. In DLS, rarely are more than three degrees of freedom found with any repeatability. Thus, with a bimodal, only the peak positions and ratio of the peak areas have meaning. The widths not so much.

This—the ill-conditioning of the general problem of deconvolution—is the fundamental limit in DLS. It prevents DLS from being a high-resolution particle sizing technique. At best, DLS is a medium resolution technique.

III.12 Various Methods for Deconvolution of the Autocorrelation Function

A variety of numerical techniques for the deconvolution have been proposed:

NNLS (Non-Negatively Constrained Least Squares)
CONTIN (one type of Regularization)
Maximum Entropy Method
Singular Value Decomposition
Exponential Sampling
Histogram Method
REPES

None are absolute as might be surmised from the limit placed on the solution by minor differences in noise from one measurement to the next. Given the limited resolution, all are usually employed with constant ratio (also called exponential) spacing on the Γ axis. And all can rarely find correct results when independent Γ values are closer than about 2:1. To be sure, finer steps can be filled in by shifting the original set of Γ values, but these are not independent of the original set. Thus, the resolution is only apparently improved. Unless the measured correlation function is very smooth (minimizing noise), the various techniques don't converge, and often result in peaks that are broader than the true distribution results. In general, none of these techniques are superior in all cases to the rest. See Chapter 4 in reference C listed near the beginning of this chapter for more detail.

III.12.1 Non-Negatively Constrained Least Squares

One technique, NNLS[30], has something in common with cumulants, William-Watts, and an assumed single exponential: linearizing the functional form (either by taking the log or the log of the log) or fitting a polynomial, none of which are ill-conditioned as far as standard least squares is concerned. This is done by starting with the discrete form for the general solution by replacing equations III-45 and III-46 with the following:

$$g^{(1)}(\tau) = \sum_{j=1}^{j=N} G_I(\Gamma_j) \cdot \exp(-\Gamma_j \tau) \tag{III-47}$$

---

[30] Grabowski, E.F. and Morrison, I.D., in *Measurement of suspended particle by quasi-elastic light scattering*, B. E. Dahneke editor, Wiley-Interscience, New York (1983).

With the usual normalization condition that the sum over the relative intensities is unity:

$$\sum_{j=1}^{j=N} G_I(\Gamma_j) = 1 \qquad \text{(III-48)}$$

Next, a numerical set of Γ_j is chosen. Usually this is done in diameter space by selecting a minimum and maximum diameter, d_{min} and d_{max}. This might be done from prior knowledge of the sample, or by calculating a d_{min} by fitting the first few channels to a single exponential. (Remember: Γ_{max} corresponds to a d_{min} and is related to the fastest part of the correlation function decay at the smallest τ values.) Somewhat more arbitrarily a d_{max}, which corresponds to a Γ_{min}, is selected usually by multiplying d_{min} by 50 or 100. After the first results are obtained, if only zeros for the $G_I(\Gamma_j)$ are found at the beginning and end of the distribution, a second pass with a larger d_{min} and smaller d_{max} can be tried. However, it is fruitless to make the spread in these initial conditions too small to get a higher resolution result. Because, what usually happens is the results don't converge or converge to a non-physical result. Again, the limit is set by the ill-conditioned nature of the deconvolution problem, and the noise along the measured function.

Though possible with very smooth correlation functions and a very accurate baseline, rarely are reliable results closer than about 2:1 in size.

Given a max and a min value for Γ, the next decision is how many values (N) should be calculated. Again, somewhat arbitrarily, a number from 10 to 50 is selected. If the number is small, a second iteration starting with a separate set of max and min Γ values is selected, amounting to a shift from the first set. This does not amount to obtaining a higher resolution result, but it does fill in results between widely spaced Γ results in the first set.

Finally, the spacing between a set of Γ_{min} and Γ_{max} is required. Early in the use of NNLS, linear spacing was used. But it was later found that this was overly optimistic. Once again, the limited resolution was better served by keeping a constant ratio between successive Γ values. This ratio can be calculated given Γ_{min}, Γ_{max}, and N. The stage is now set for the calculation. (Ratio spacing of results is also the reason that ratio spacing of correlator delay time values τ is available in advanced designs.)

This time, however, the parameters to be fit, the set of $G_I(\Gamma_j)$ are linearly related to $g^{(1)}(\tau)$ since the term $\exp(-\Gamma_j\tau)$ is just a number at each delay time τ and each one of the N values of Γ_j. Thus, the ill-conditioning has been lifted by this linearization trick.

Still, without using the non-negativity condition—the NN in NNLS—perfectly good fits can be obtained with some of the $G_I(\Gamma_j)$ being negative. But since these represent relative intensity values, which are inherently positive, adding the NN condition constrains the solution to a physically realistic set of positive $G_I(\Gamma_j)$. An example is shown for a bimodal in Figure III-7 using NNLS.

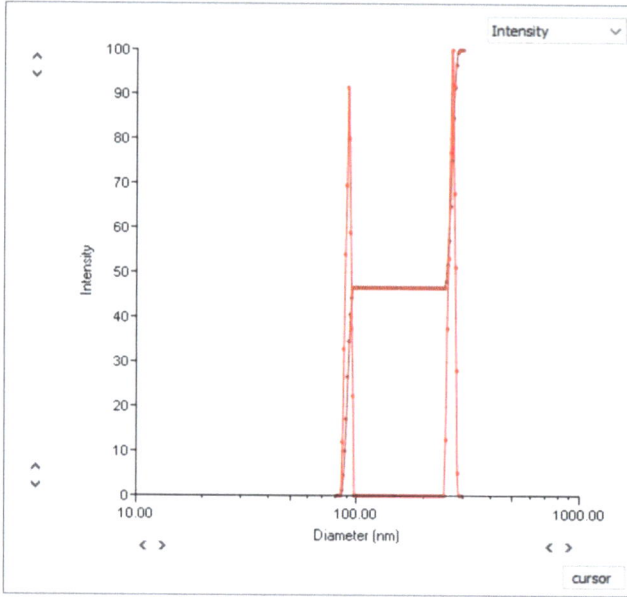

Figure III-7: Bimodal size distribution using NNLS.

The sample was a 50:50 mixture by intensity of (92 ± 2) nm polystyrene (PS) latex in 10 mM KCl plus (269 ± 5) nm PS. Cumulant results show an Effective Diameter of 143 nm and a Polydispersity Index of 0.199. Therefore, the distribution is not monodispersed. Tabular data for the NNLS result shows one peak at 90.8 nm and the other at 264 nm, with 47% by intensity in the smaller peak and 53% in the larger. Notice the peaks are narrow on this log scale. The fit agrees within experimental error of the known, mixed distribution.

Compare this to the CONTIN result on the same raw correlation function shown in Figure III-8.

III.12.2 CONTIN (One form of Regularization)

As mentioned, one problem of ill-conditioning is that artifacts in the form of pseudo peaks can result, though they are usually not repeatable. One way around that is to add another term to the sum of the square of the residuals to smooth results and penalize spurious peaks.[31] See equation III-49:

$$S = \sum_i w_i \cdot \left(g^{(1)}(\tau_i) - \sum_j G_I\left(\Gamma_j\right) \cdot exp\left(-\Gamma_j \cdot \tau_i\right) \right)^2 + \alpha \cdot f^{reg} \quad \text{(III-49)}$$

The factor α is called the regularization parameter. It varies from 0 to 1. When it is zero, the result is the same as in equation III-44. When it is 1, the smoothing term is as important as the minimization of the sum of the squares of the residuals. Over smoothing flattens peaks and makes it harder to separate them. Thus, compared to NNLS, CONTIN is best used on

---

[31] Tikhonov, A.N, *Soc. Math*, **4**, 1035 (1963)

broad unimodal distributions (polymers, for example) or widely separated peaks such as 3:1 and higher.

Different forms for the regularization term f^{reg} are possible. The most common is the square of the sum of the 2nd derivative of $G(\Gamma_j)$. A program called CONTIN uses this method but allows for other choices instead of the 2nd derivative.[32] For each α, a different distribution is obtained. It is up to the user to determine which is appropriate. The default value for α is 0.5. But for certain samples 0.3 or 0.7 produce better results. Of course, that requires *a priori* information about the shape of the distribution.

An example is shown for a bimodal in Figure III-8 using CONTIN. The same correlation function was used from the same 92 nm-269 nm mixture of PS latex in 10 mM KCl. The same equal intensity mixture was used. The cumulant results will be, of course, the same.

The 1st peak is at 91.5 nm with 47% by intensity and the 2nd peak is at 262 nm with 53% by intensity. Clearly, these results are consistent with expectations and in excellent agreement with NNLS results.

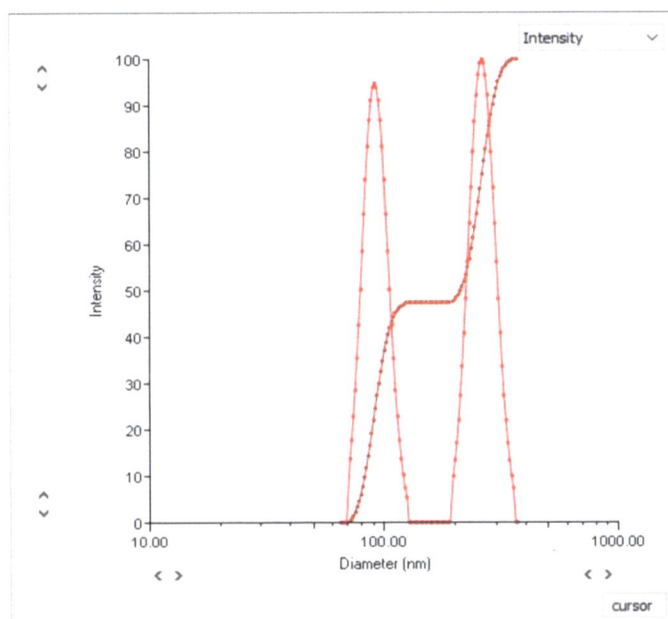

Figure III-8: Bimodal size distribution using CONTIN.

---

[32] Provencher, S.W., *Computer Phys. Comm.*, **27**, 213, (1982).

II.13 Summary Comments on Size Distributions using DLS

Before extracting a size distribution, make sure the measured correlation function is smooth and the baseline is flat. If not, getting more information than an average size and a polydispersity index is problematic. Distribution results will not be repeatable. Still, getting the first two moments to be repeatable is often enough to characterize the distribution.

Peaks closer than 2:1 apart are suspicious and a few repeats should be attempted. A trimodal is almost always wrong, with one or more of the peaks being artifacts.

Problems with the baseline are apt to yield large diameter artifacts in the results. Recalculate with a different baseline to see if the large diameter peak appears again. Problems with high frequency noise along the correlation function are apt to yield small diameter artifacts in the results. Increase measurement duration, or laser power to see if they persist.

If two or more algorithms yield consistent results such as shown in Figures III-7 and III-8, then results are more believable. With multi-modals, if one peak's diameter is large enough that it will show an angular dependence, then make another measurement at a higher angle. The peaks should appear again, but the relative intensity (area under the curve in the differential distribution) should decrease at the higher angle for the larger particle.

Finally, compare the distribution with any other information you have. Is it consistent with other types of measurements including SLS results? Latex particles may aggregate but they don't usually fall apart. Thus, the appearance of a small-diameter peak is probably an artifact.

Chapter IV: Microrheology Measurements using DLS

IV.1 Introduction

Rheology is the study of elastic deformation and viscous flow. Energy is either stored elastically or dissipated viscously. In one extreme, a perfectly elastic solid stores energy when it is deformed. The viscous response is insignificant if not zero. In another extreme, a simple liquid, only the viscous response is significant, and the elastic response is zero. Complex fluids may exhibit both such properties. The complexity arises from internal structure due to dissolved polymers, natural and synthetic, and any coupling (networks) between particles and molecules that constitute the dispersed phase.

Microrheology is the study of these viscoelastic properties using "micro" probes such as a submicron tracer (also called probe) particle. Classical rheology uses mechanical rotational devices and as such is limited to volumes of more than a few milliliters and to frequencies below about 100 Hz = 6,000 rpm = 628 rad/s. Using DLS on a dilute complex fluid, the frequency range is extended by several orders of magnitude and the sample volume can reduced to microliters.

The result of such a study is the viscoelastic parameters η^*, G" and G': the complex viscosity, the viscous loss modulus, and the elastic storage modulus, respectively.

Since the probe particles' motion will not be forced, but responds to thermally driven forces, this type of microrheology is called passive. And linear rheological properties are the result. Passive microrheology is for use in weakly interacting complex fluids that are dominated by viscous forces, so we expect the elastic moduli to be insignificant compared to the viscous moduli.

Finally, we will use DLS to monitor the probe particles' motion. Thus, we will work in dilute systems that look either apparently clear or slightly turbid but with no multiple scattering. The complex viscosity, analogous to the normal viscosity of a simple liquid (Newtonian liquid), will be relatively small, up to a few mPa·s (note 1 mPa·s = 1 cP).

IV.2 Samples and Criteria for Success

The underlying structures in the liquid, caused by dissolved polymers or proteins, for example, also scatter light. They will also affect the shape of the measured autocorrelation function. Thus, use a high enough probe concentration to dominate the scattering. If uncertain, make measurements at three probe concentrations and choose the lowest concentration of the two where the shape of the rheological properties does not significantly change. Why the lowest of the two? The highest concentration of the two may suffer from multiple scattering causing changes in the shape of the ACF that has nothing to do with microrheology.

If in doubt, measure the ACF of the probe particle at the same concentration in the simple fluid without the underlying structure (if possible). Since the size of the probe should be known, if there is significant multiple scattering, the calculated size is always smaller indicating multiple scattering.

The size of the probe particle should be larger than the microstructure of the complex fluid. This is easier said than done, since there is no simple formula to determine length scale of the microstructure. As a first guess, estimate or measure R_H, the hydrodynamic radius from DLS, of the sample without the probe. True, since the correct viscosity is not known, the estimate of the hydrodynamic radius is a bit in error. Ignore this and start with a probe radius at least a few times larger than this R_H. If possible, make measurements at two or three difference probe sizes and see if the microrheological results are constant as they should be when the criterion is fulfilled.

The probe should not interact with the structure other than to indicate viscous loss and elastic storage. An example of interaction might be that the charges on the probe particle interact with charges (if any) on the molecules causing the structure. For instance, a negatively charged polyelectrolyte causing structure might influence the movement of a positively charged probe particle. A rough estimate of any interaction is the zeta potential of the probe particle in the simple liquid and again in the simple liquid plus the molecules that give rise to the structure. Now the zeta potential depends on the viscosity and dielectric constant of the liquid and so this might be hard to estimate for the complex fluid. In this case, use the same values as those used to calculate the results in the pure liquid. If the zeta potentials don't differ by more than an estimated 20% or so, it is okay to assume the probe and structure are not interacting.

Sample preparation takes a bit of patience. In making the samples whose results are shown below, the polymers were dissolved thoroughly using rolling, often overnight. When the probe particles were added, rolling is required for several more hours to ensure the probe particles are distributed randomly throughout the dissolved polymer samples.

Likewise, when temperature measurements were made, samples equilibrated for at least 30 minutes after the oven reached the designated temperature.

IV.3 Making Measurements

This is much like any DLS measurement: A measured autocorrelation function, after normalized by the baseline, is analyzed to yield parameters of interest. However, in microrheology, the size is not calculated; it is an input. It is a single value, so use a monodisperse sample. It should be large: a radius of 30 nm for Newtonian liquids; but start with 200 nm and larger for visco-elastic fluids. In the examples below, 496 nm and 410 nm-radius probe particles were used.

IV.4 Microrheological Properties: The Outputs of a Measurement

The properties of interest are the graphs that result from the calculations performed on the measured autocorrelation function. Here are the choices:

The *complex viscosity*, $\eta*$ in mPa·s (cP) vs. ω in rad/s: If this curve is horizontal and does not change with frequency, then the single value of $\eta*$ is equal to the viscosity of this Newtonian liquid at the temperature of the measurement. If the curve decreases when ω increases, then this is called *shear thinning*. The liquid becomes less viscous the higher the frequency. If the curve increases when ω increases, then this is called *shear thickening*. The liquid becomes more viscous the higher the frequency. Shear thinning is more common than shear thickening.

The *viscous loss modulus*, G" in Pa vs. ω in rad/s: Generally, this curve rises when ω increases. Energy is dissipated more at higher frequencies through viscous loss. This is the more common case with the dilute complex fluids examined by DLS.

The *elastic storage modulus*, G' in Pa vs. ω in rad/s: Generally, this curve rises when ω increases. Energy is stored more at higher frequencies; however, in the case of the dilute complex fluids examined by DLS, G' is less than G", often by an order of magnitude or more.

It is common to overlay G" and G' vs. ω.

The *Mean-Square-Displacement, MSD in nm² vs. τ, delay time in μs:* Finally, to emphasize that the calculations are derived by examining how the complex structure of the fluid hinders the otherwise free diffusion of the probe particle, the Mean-Square-Displacement of the probe vs. delay time is commonly plotted.

Please note that neither a size nor size distribution is output. Size is an input: the probe radius. It is not an output. Additionally, it is the shape of the graphs that are the outputs and how they change with temperature, concentration of underlying structure (polymer concentration, for example), and structure size (MW of polymer determines chain size and therefore structure size). Sections of these Log-Log graphs are often straight lines and their slopes characterize the type of structure present.

IV.4.1 Examples of Outputs

Here is a series of outputs. The sample was 150 kDa Dextran, a water-soluble polymer, 2 mg/mL (approx.) with 993 nm diameter polystyrene latex as the tracer, so a probe radius of 496 nm.

Noise at low frequency can be traced to the slight differences between the correlation function and the baseline at high delay times, and the noisy shapes of the curves should therefore be ignored. This applies below to η*, G" and G' at low frequency and to the MSD at high delay time.

Subtle wiggles in the curves may be due to calculation errors arising from the relatively few correlation points and their spacing, but also from the fact that these polymers are not narrowly distributed. Thus, different chain lengths may respond differently with frequency. Finally, at short delay times, corresponding to high frequencies, if the scattering from the structure is not completely overshadowed by that from the probe particles, there may be some evidence in the autocorrelation function that appears as wiggles in the microrheological curves. It is better to make measurements at a series of probe particle concentrations and compare the effects.

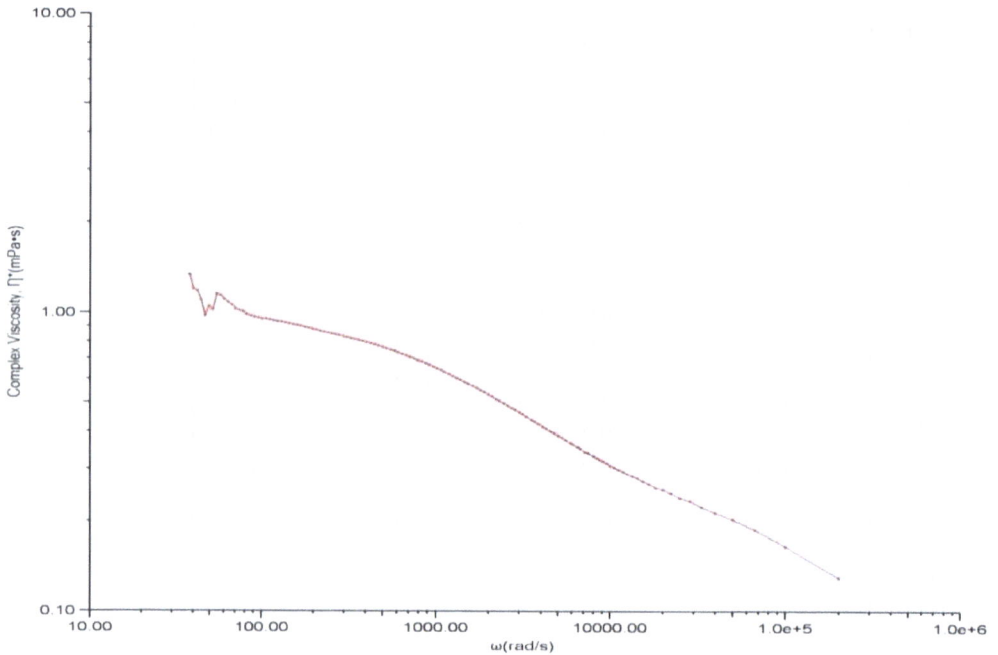

Figure IV-1: Complex viscosity for 150 kDa Dextran, 2 mg/mL, at 25 °C.

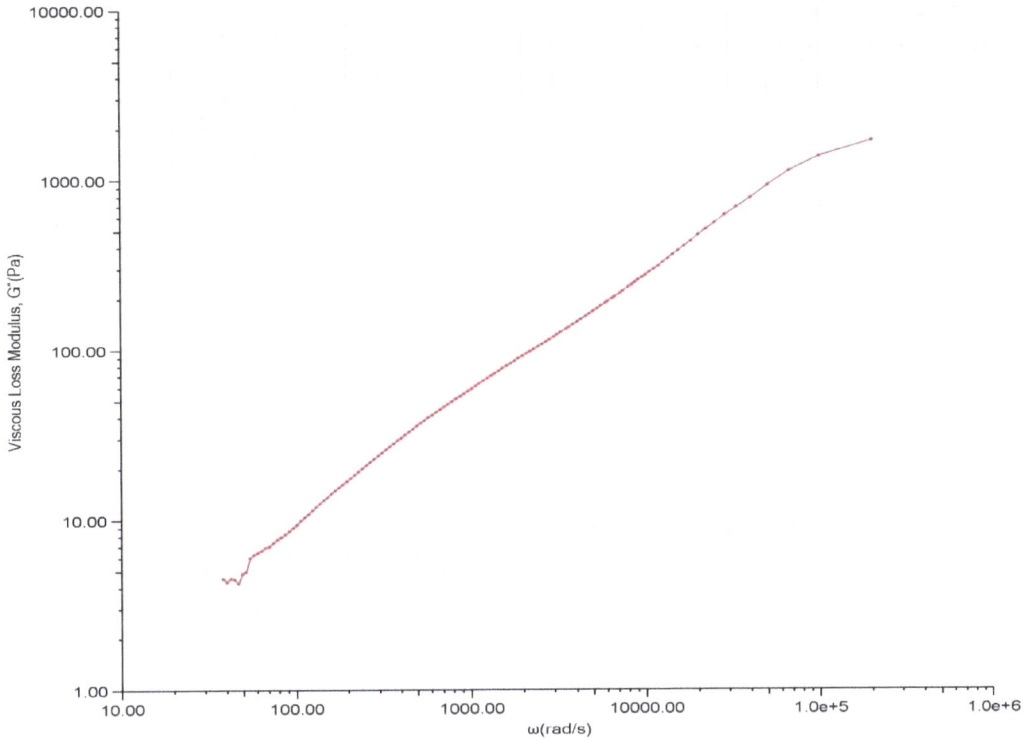

Figure IV-2: Viscous loss modulus for 150 kDa Dextran, 2 mg/mL, at 25 °C.

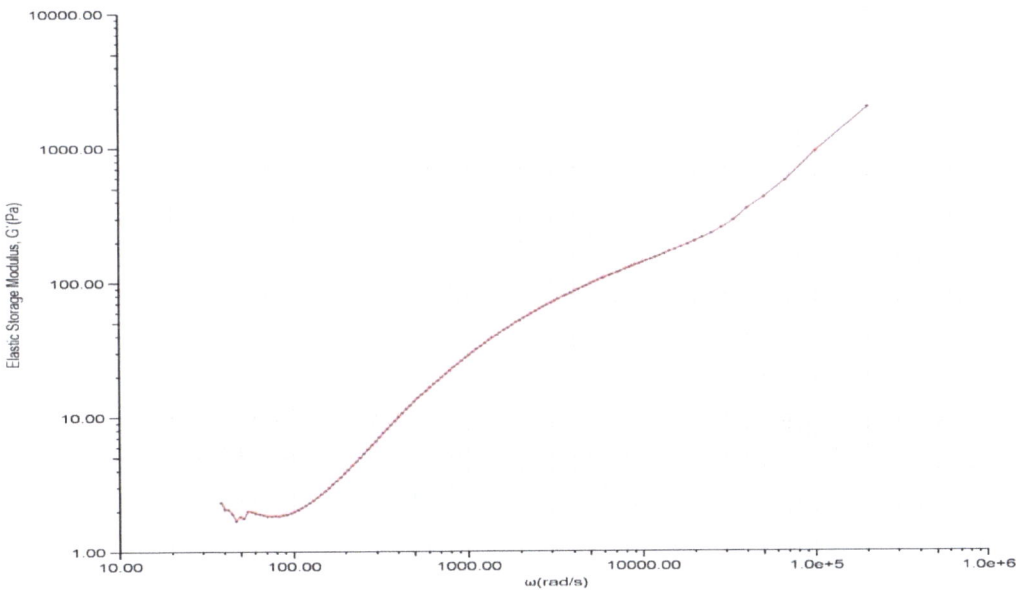

Figure IV-3: Elastic storage modulus for 150 kDa Dextran, 2 mg/mL, at 25 °C.

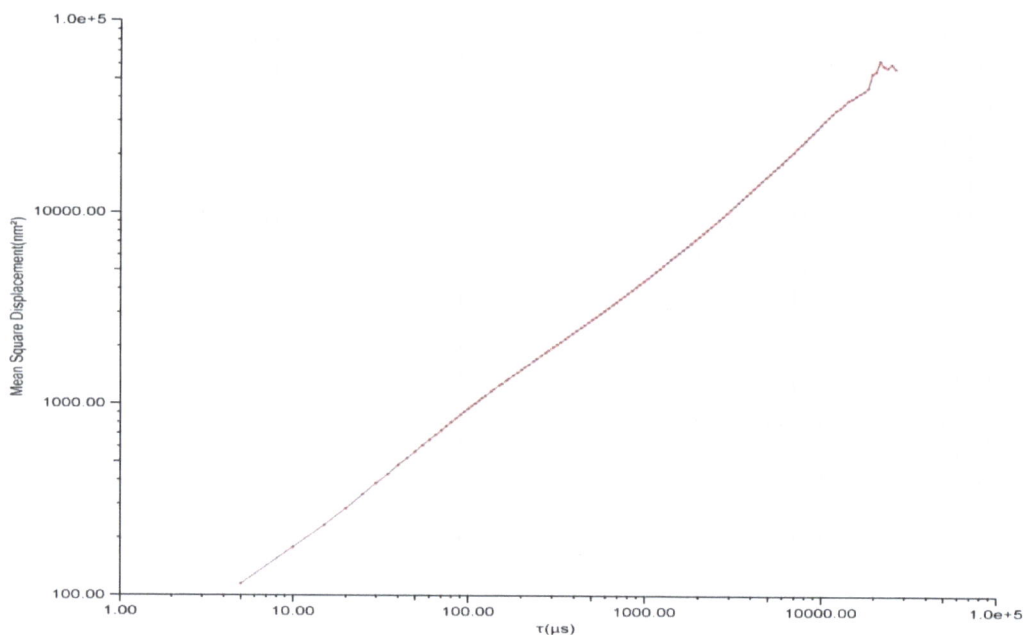

Figure IV-4: Mean square displacement of 993 nm diameter probe of polystyrene sphere in 150 kDa Dextran, 2 mg/mL, at 25 °C.

Temperature can cause a difference in the structure and therefore in microrheological properties. Here is an example of G", the viscous loss modulus, at 25 °C and 45 °C. The sample was 0.54 mg/mL polyvinylpyrrolidone, PVP, at a nominal molecular weight of 10 kDa. Note that this sample is somewhat polydisperse and the molecular weight is M_w determined by GPC. The tracer particle is a polystyrene latex sphere of nominal diameter 820 nm stabilized by sodium dodecylbenzene sulfonate. The concentration of the tracer particle was 10 µL of a 5% (50 mg/mL) w/v suspension in 20 mL of the dissolved polymer, so 25 µg/mL.

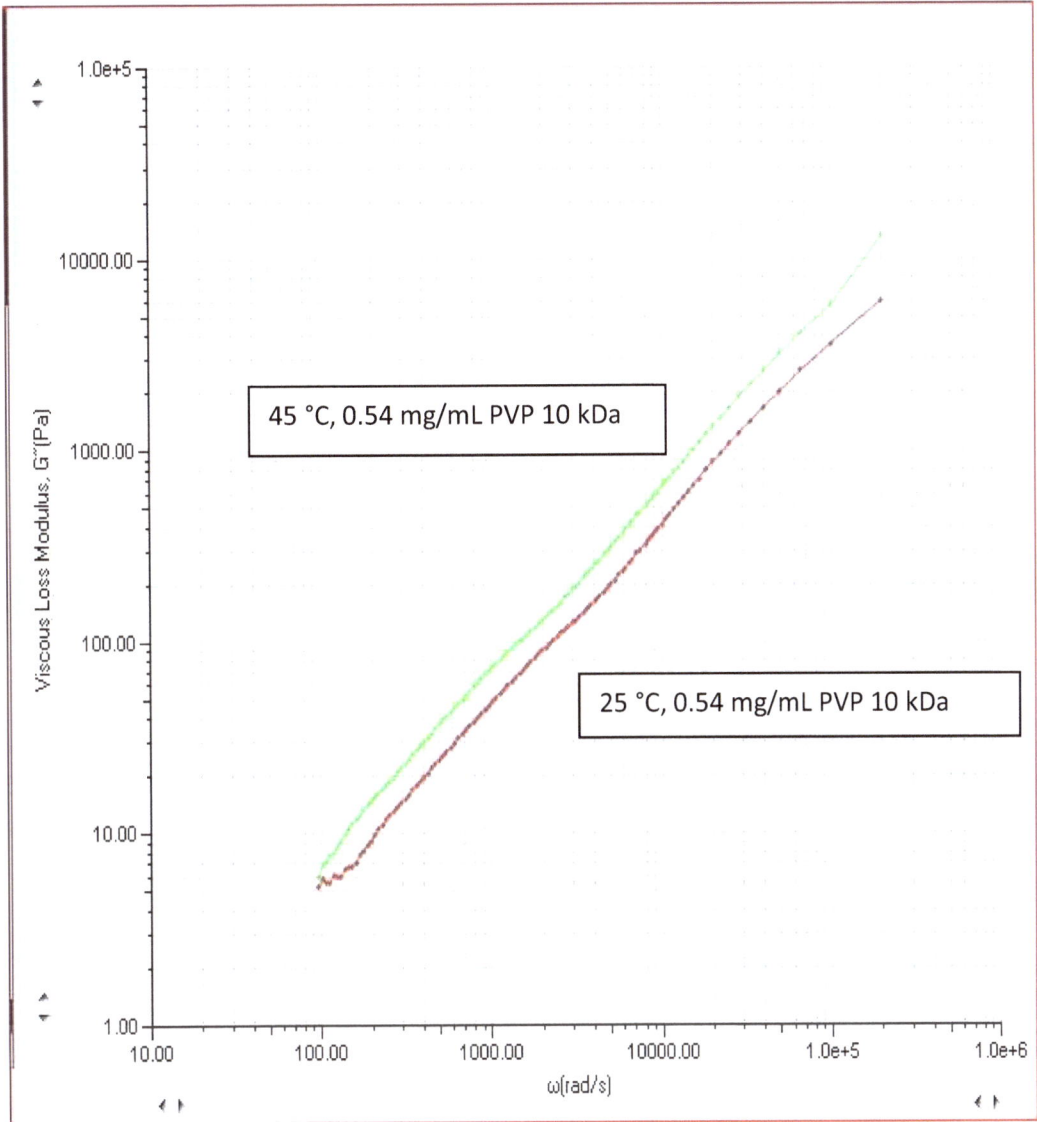

Figure IV-5: Viscous loss modulus at 25 ℃ and 45 ℃ using 10 kDa PVP at 0.54 mg/mL.

Polymer concentration can cause a difference in the structure and therefore in microrheological properties. Here is an example of η^*, the complex viscosity, at 2 mg/mL and 10 mg/mL of polyvinylpyrrolidone at a nominal molecular weight of 40 kDa. Note that this sample is somewhat polydisperse and the molecular weight is M_w determined by GPC. The tracer particle is a polystyrene latex sphere of nominal diameter 820 nm stabilized by sodium dodecylbenzene sulfonate. The concentration of the tracer particle was 10 μL of a 5% (50 mg/mL) w/v suspension in 20 mL of the dissolved polymer, so 25 μg/mL.

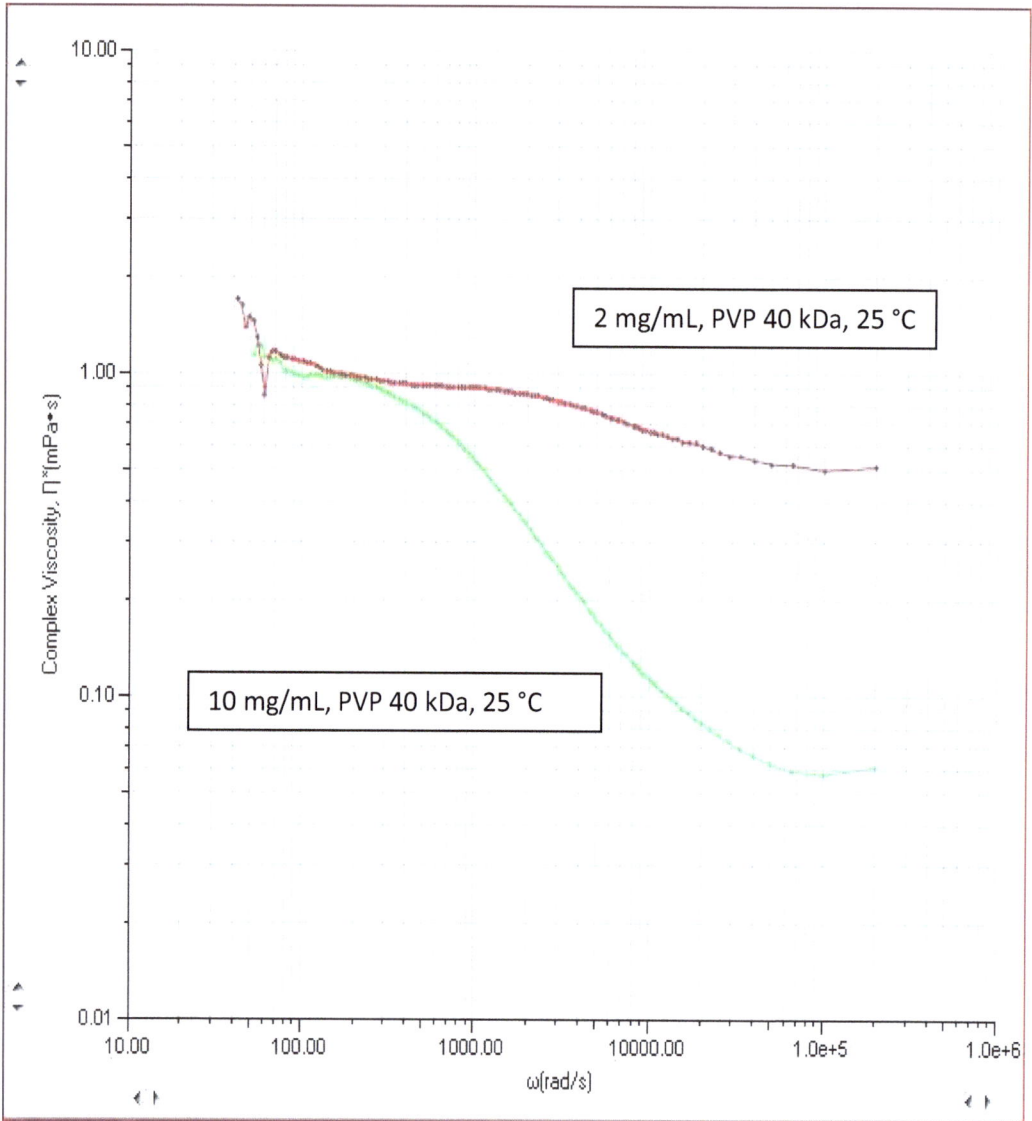

Figure IV-6: Complex viscosity at 25 °C of 40 kDa PVP at 10 mg/mL and 2 mg/mL.

Both curves display shear thinning. The difference is that the 10 mg/mL sample shows an increasing amount beyond about 300 rad/s, a range which cannot be detected by mechanical rheometers.

IV.5 Theory of DLS Microrheology

Starting with DLS in a simple (Newtonian fluid), using self-beating, on a monodisperse, spherical particle, with translational diffusion coefficient D_T, we know the measured, un-normalized, intensity (also called the 2nd order) autocorrelation function looks like this:

$$G_2(\tau) = B[1 + G_2(0)e^{-2\Gamma\tau}] \qquad \text{(IV-1)}$$

Dividing by the baseline B, subtracting 1, and taking the square root yields the normalized, electric field autocorrelation function:

$$g_1(\tau) = g_1(0)e^{-\Gamma\tau}, \quad \text{where} \tag{IV-2}$$

$$\Gamma = D_T q^2, \quad \text{with} \tag{IV-3}$$

$$D_T = k_B T / 6\pi\eta a, \quad \text{and} \tag{IV-4}$$

$$q = \frac{4\pi n}{\lambda} \cdot \sin\frac{\theta}{2} \tag{IV-5}$$

Here the symbols have their usual meanings:

θ = Scattering angle
λ = Laser wavelength
N = Liquid refractive index
a = Tracer (probe) particle radius
η = Liquid viscosity
T = Absolute temperature (K)
k_B = Boltzmann's constant
D_T = Translational diffusion coefficient
Γ = Linewidth of the power spectral density function
τ = Delay time as determined by the autocorrelator
$G_2(\tau)$ = Un-normalized, intensity ACF coefficient at each delay time
$G_2(0)$ = Un-normalized, intensity ACF coefficient at zero delay time
$g_1(\tau)$ = Normalized, electric field ACF coefficient at each delay time
$g_1(0)$ = Normalized, electric field ACF coefficient at zero delay time

Equation IV-4 is called the Stokes-Einstein equation and it relates diffusive motion to liquid viscosity (and temperature) and particle size assuming there is only viscous loss. This is the definition of a simple or Newtonian liquid.

In the early part of the 20th century, Einstein related the mean-square-displacement, MSD, of a particle undergoing thermal motion to its translational diffusion coefficient using random walk ideas:

$$MSD = \langle (r(\tau) - r(0))^2 \rangle = \; <(\Delta r)^2> \; = 2 \cdot d \cdot D_T \cdot \tau, \quad \text{where} \tag{IV-6}$$

d = 1 for one dimensional motion, 2 for motion in a plane, and 3 for three-dimensional motion (the normal kind measured with particles in a fluid). Here r is a distance moved by the particle.

Substituting this and equation IV-3 into equation IV-2 relates the MSD to the measured, normalized, 1st order autocorrelation function:

$$g_1(\tau) = g_1(0)e^{-q^2 \cdot <(\Delta r(\tau))^2>/6} \qquad \text{(IV-7)}$$

With complex fluids, where viscous and elastic features play a role, the Langevin equation was used to produce a _Generalized_ Stokes-Einstein equation.[33] One way of writing it is this:

$$G(\omega) = \frac{k_B T}{\pi a <(\Delta r(1/\omega)^2 > \Gamma[1+\alpha(\omega)]} \qquad \text{(IV-8)}$$

Notice how this resembles equation IV-4, the Stokes-Einstein equation. Here Γ is the gamma function, a mathematical function, not to be confused with Γ the linewidth; α is a variable with limits of 1 for pure viscous and 0 for pure elastic motion; and $\omega = 1/\tau$.

The theory further shows that the viscous loss modulus, G", is given by:

$$G" = G(\omega) \cdot \sin[\alpha(\omega)\pi/2] \qquad \text{(IV-9)}$$

And that the elastic storage modulus, G', is given by:

$$G' = G(\omega) \cdot \cos[\alpha(\omega)\pi/2] \qquad \text{(IV-10)}$$

When $\alpha = 1$, the cosine of $\pi/2$ is zero and this ensures the elastic storage modulus is zero, leaving only G", a purely viscous system. When $\alpha = 0$, the $\sin(\pi/2)$ is zero and this ensures the viscous loss modulus is zero, leaving only G', a purely elastic system. Where DLS microrheology applies in dilute solutions with weak structure, the elasticity is small, and the viscous losses dominate so α will have values closer to unity than to zero.

Finally, G" and G' can be combined to yield the complex viscosity, η^*, as given by:

$$\eta^* = \sqrt{\frac{[G"(\omega)]^2 + [G'(\omega)]^2}{\omega^2}} \qquad \text{(IV-11)}$$

IV.6 Data Analysis

The goal is to calculate G(ω) from which G", G', and η^* are obtained. To calculate G(ω), start by taking the natural logarithm of equation IV-7 and rearranging:

---

[33] Mason TG, Weitz DA, Phys. Rev. Lett., 74, 1250-1253, 2000

$$< \left(\Delta r(\tau) \right)^2 >= \frac{6}{q^2} [Ln(g_1(\tau)) - Ln(g_1(0))] \qquad \text{(IV-12)}$$

Before this can be substituted into equation IV-8, one needs $g_1(0)$, the normalized, electric field autocorrelation coefficient at zero delay time. This can't be measured directly but can be obtained by fitting $Ln(g_1(\tau))$ vs. $Ln(\tau)$ and extrapolating to zero. So, from the normalized, electric field autocorrelation function, $g_1(\tau)$, the intercept $Ln(g_1(0))$ is obtained. The remaining variable is $\alpha(\omega)$.

For a long time, this was difficult to calculate from DLS data because the raw data is not equally spaced in delay time; however, Mason[34] found a way to do it. It amounts to plotting Ln(MSD) vs. Ln(τ) and determining the slope at each point:

$$\alpha(\omega) = \frac{dLn<(\Delta r(\tau))^2>}{dLn(\tau)} \qquad \text{(IV-13)}$$

Summary: From the normalized, electric field autocorrelation function, plot $Ln(g_1(\tau))$ vs. $Ln(\tau)$. Extrapolate to zero to obtain $Ln(g_1(0))$. Given this value, use equation IV-12 to obtain the MSD. Plot Ln(MSD) vs. Ln(τ) and the slope at each point yields α(ω). Use the MSD and α(ω) as well as the probe particle radius "a" in equation IV-8 to obtain G(ω) as a function of ω. Use equations IV-9, IV-10 and V-11 to obtain G", G', and η* as functions of frequency ω.

IV.6.1 Advanced Data Analysis

If the measured ACF is not smooth and the baseline not flat, do not proceed. Nothing good will come of it. Make a new sample if necessary and make a new measurement. The MSD depends on finding an excellent value of the log of the intercept, $Ln(g_1(0))$. And the MSD is crucial in determining α(ω). Given the discrete nature of the raw data and its nonlinear spacing, uncertainties in α(ω) arise. Any value of α(ω) > 1 is not physically possible. But simply making all values less than or equal to 1 is insufficient. Equation IV-10 shows that a value of 1 or even values ever so slightly less than 1, combine with the cosine function to cause sharp dips in G' which are not real. For these reasons scan the list of alpha values and set any higher than, say, 0.98 equal to 0.98. Often, even that fix is not enough and using multipoint averaging to smooth alpha further is required.

Maximum Alpha: It is a value from 0.99 to 0.90. By default, set it to 0.98. If G' is important to your work and it displays sharp declines followed by sharp rises and repeats of this kind but not with any regular frequency, try a lower value like 0.97 or 0.96.

---

[34] T.G. Mason, Rheol. Acta, 39, 371-378, 2000

Alpha Smoothing: The variation in the point-to-point values in $\alpha(\omega)$ is a direct result of the non-linear spacing of the correlation function and the method embodied in equation IV-13 to obtain a derivative. As a further attempt to damp oscillations, multipoint smoothing is suggested. Odd values from 3 to 11 may be used. By default, use 7-point smoothing. It works like this: Add the values of alpha for three points before, three points after, plus the point of interest. Divide by seven and use that for alpha at the point of interest.

Different smoothing makes a small difference in most cases. Choose the value that results in the smoothest final graphs of interest.

Intercept Factor: It is a multiplier of $Ln(g_1(0))$ used to determine MSD. The derivative of MSD is used to determine the alpha values. The intercept may be affected by the noise on the autocorrelation function and your choice of baseline. In any case, allowing for a change in $Ln(g_1(0))$ by using a multiplier, allows the user to see what a variation will do to the end results, that is to the shape of the various graphs.

By default, the multiplier should be 1. Values from 1.05 to 0.95 are suggested with those closest to 1 being recommended. Using such a multiplier can change most dramatically the tails (moving them up or down) of the curves at high ω (low delay time, τ, so effects the MSD graph at low delay times) because the intercept is most important in that range. G' is affected the most.

Intercept Polynomial Degree: Consider using the first 20 normalized points of the auto-correlation function and fit to either a 1st, 2nd or 3rd order polynomial. The point is to find a more accurate $Ln(g_1(0))$. Of the four variables allowed in advanced data analysis, this one is the least important. The default value of 3rd order is usually enough. After all, a change in the intercept can be more easily controlled using the Intercept Factor.

IV.7 Summary

It is curious that light scattering yields visco-elastic information on complex fluids. However, as probe particles are hindered in their diffusive motion by structure in the liquid, the link between the stress and strain in that thermally derived motion and dynamic light scattering is a little more obvious. DLS, after all, is probing that same motion.

Chapter V: Electrophoretic & Phase Analysis Light Scattering: μ_e, ζ

V.1 Optical mixing (Heterodyning) and Electrophoretic Light Scattering ELS)

What happens if the detector, in addition to the scattered light, E_s, picks up a little bit of light from the same source called a reference beam or local oscillator, E_{LO}. Unlike stray or flare light, which are uncontrolled, the controlled reference beam leads to an interesting result. Starting with equation III-19, the result is:

$$G^{(2)}(\tau) = B\left(1 + f^2 \left(\frac{E_s}{E_T} exp(-\Gamma\tau) + \frac{E_{LO}}{E_T}\right)^2\right) \qquad \text{(V-1)}$$

Where the effect of a reference beam has been added to the usual electric field ACF. The total electric field $E_T = E_s + E_{LO}$. In equation III-19, the scattered field didn't appear because, when $E_{LO} = 0$, $Es/(Es + 0) = 1$. In the other extreme, when the local oscillator is much stronger than the scattered light, so $E_{LO} >> E_s$, the time dependent term is insignificant, and the correlation function doesn't appear to decay. Imagine now the reference beam is controlled so the cross term dominates:

$$G^{(2)}(\tau) \approx B\left(1 + f^2 \left(\frac{2E_s E_{LO}}{E_T^2} exp(-\Gamma\tau) + \frac{I_{LO}}{I_T}\right)\right) \qquad \text{(V-2)}$$

In this case, $E_{LO} > E_s$, but not very much greater than. By controlling the intensity of the reference beam, it is possible to isolate the correlation function of the scattered electric field.

And, as in equation III-11, the power spectral density function is given by:

$$P_E(\omega) = \frac{1}{\pi} Re \int_0^\infty e^{-\Gamma\tau} e^{-i\omega\tau} d\tau = f^2 \left(\frac{2E_s E_{LO}}{E_T^2}\right) \frac{\Gamma/\pi}{\Gamma^2 + \omega^2} \qquad \text{(V-3)}$$

again, ignoring the delta function at zero frequency and the constant shot noise term.

It is plotted in Figure III-4. Now suppose a constant value ω_o is added to the reference beam. In this case, the result is given by:

$$P_E(\omega) = f^2 \left(\frac{2E_S E_{LO}}{E_T{}^2} \right) \frac{\Gamma/\pi}{\Gamma^2 + (\omega - \omega_o)^2} \qquad \text{(V-4)}$$

This is a shifted Lorentzian and is shown in Figure V-1. The half-width at half-height (maximum) is still Γ. But it is shifted away from $\omega = 0$. It is now centered at ω_o.

If there is a further shift due to directed motion of the particle, it will now be with respect to the reference frequency, not zero. This allows determination of positive and negative shifts. Without the reference beam, a negative shift would not be measureable.

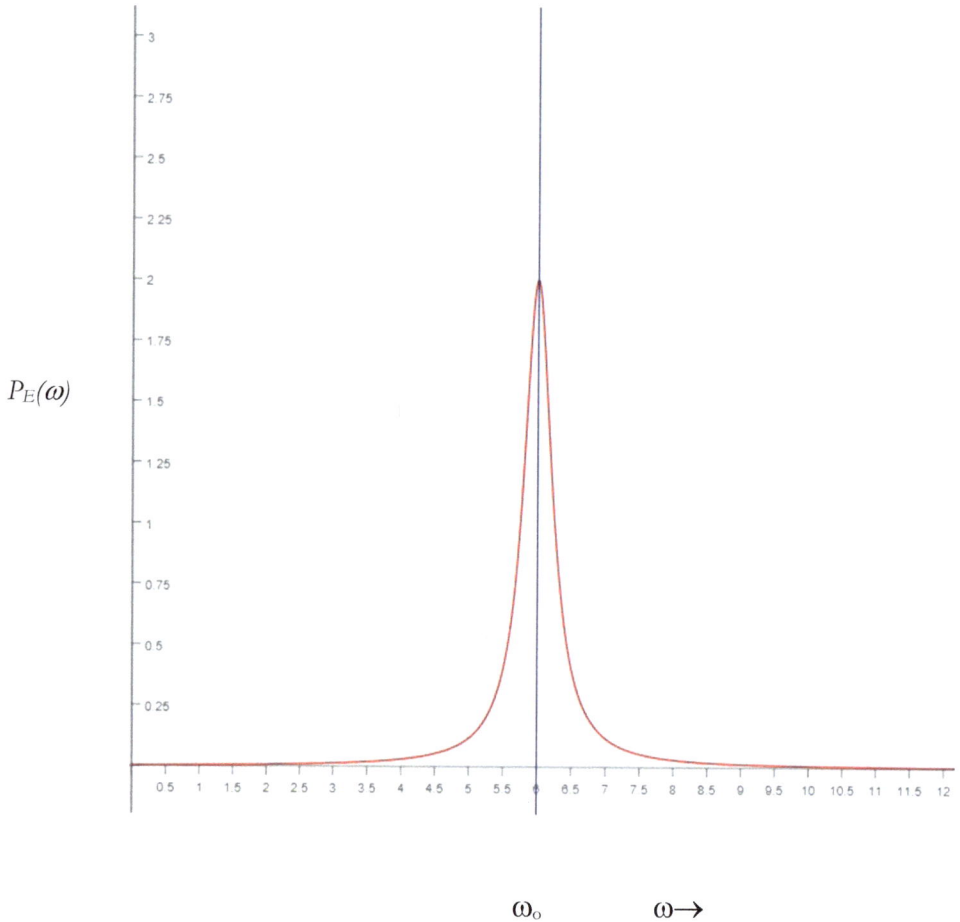

Figure V-1: A Lorentzian shifted by a reference beam, a local oscillator with added frequency shift of ω_o.

Charged particles in an electric field move with an electrophoretic velocity V_e given by:

$$V_e = \mu_e \cdot E \qquad \text{(V-5)}$$

where both velocity and field are shown in bold to designate vectors. The proportionality constant defines the electrophoretic mobility, μe, from which zeta potential is calculated.

The electrophoretic velocity gives rise to a Doppler frequency shift, ω_s, in the power spectral density function where:

$$\omega_s = \boldsymbol{V_e} \cdot \boldsymbol{q}$$

(V-6)

Once again, the scattering wave vector is designated by q. And the Doppler shifted frequency is a vector dot product. The power spectral density distribution is now given by:

$$P_E(\omega) = f^2 \left(\frac{2E_s E_{LO}}{E_T{}^2} \right) \frac{\Gamma/\pi}{\Gamma^2 + [\omega - (\omega_o \pm \omega_s)]^2}$$

(V-7)

This too is a shifted Lorentzian, shifted to either side of the reference beam frequency. For a positive shift, $+\omega$s, the result is shown in Figure V-2:

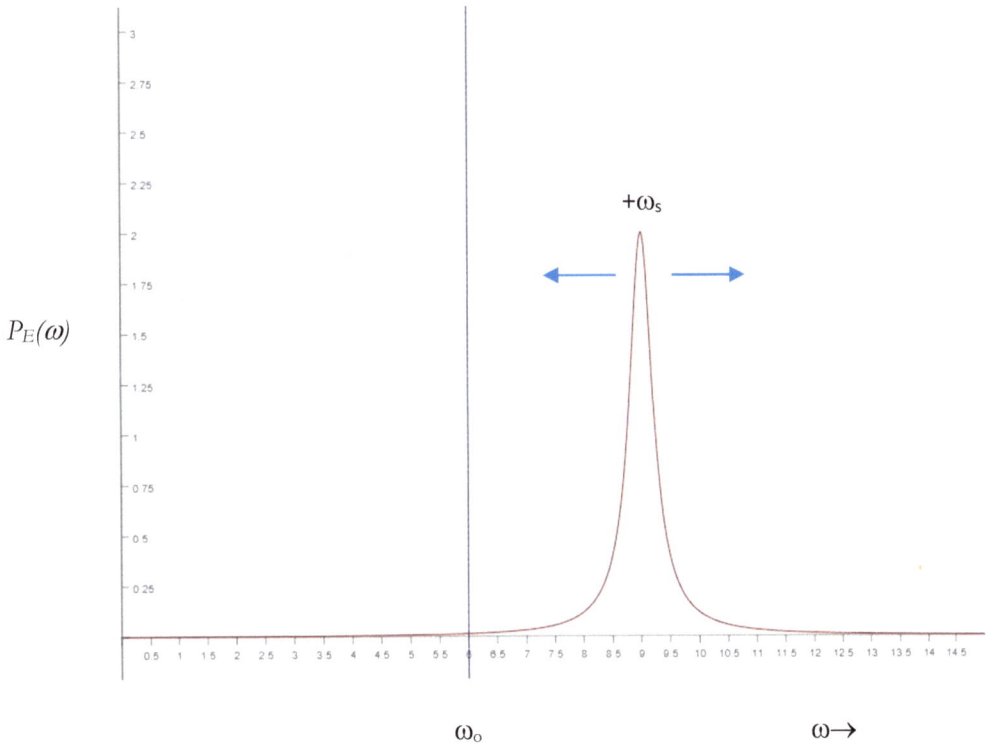

Figure V-2: A Lorentzian shifted by a reference beam, a local oscillator with added frequency shift of ω_o, and a positive Doppler frequency shift ω_s. Note: A negative Doppler frequency shift would appear to the left of ω_o.

In summary, electrophoretic light scattering proceeds as follows: The power spectral density distribution P_E is measured, and the Doppler shifted frequency ω_s on either side of the reference frequency ω_o is determined. The electrophoretic velocity V_e is calculated using ω_s and q in equation V-6. The electrophoretic mobility μ_e is calculated from V_e and the applied electric field E in equationV-5. Finally, the zeta potential ζ is calculated from μ_e using one of a few different equations. For more information consult the book by Johnson and Gabriel.[35]

Figure V-3 shows an example of a real ELS measurement on an organic pigment in 1 mM KNO_3.

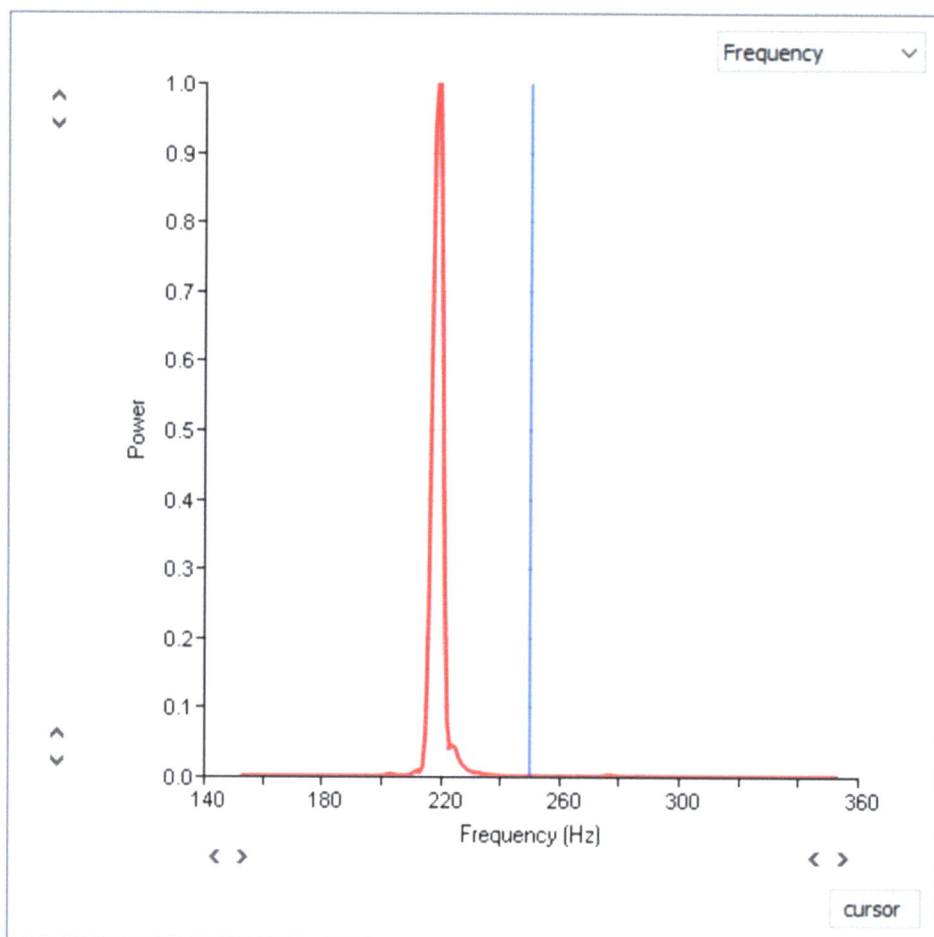

Figure V-3: Real ELS measurement. The reference frequency is at $\omega_o = 250$ Hz (blue line) and the shift to ~ 220 Hz means ω_s ~ - 30 Hz, from which a negative zeta potential can be calculated.

---

[35] Johnson, C. S. Jr. and Gabriel, D., "Laser Light Scattering", Chap. 3.C.3, Dover, 1981.

The calculation of ζ from μ_e is explored in detail later in the chapter. But, first, it is important to explore some colloid fundamentals including the definition of zeta potential.

V.2 Colloid Fundamentals

Remember that the colloidal classification includes nanoparticles, proteins, other biomacromolecules, and polymers. A brief, simplified description of the solid-liquid interface follows. It also applies with a little modification to liquid-liquid and liquid-gas interfaces. The reader should not assume this description is complete. On the contrary, these few pages constitute nothing more than a sketchy outline, just enough to understand what is being measured and why zeta potential is significant. For a deeper understanding, please see Hunter's book.[36]

Almost all solids acquire a surface charge when placed in polar liquids. This is the rule, not the exception. The charge can arise in several diverse ways. Among the more common are:

- Ionization of surface groups such as carboxyl and amino groups
- Adsorption of ions such as surfactant, multivalent ions, and polyelectrolytes
- Unequal dissolution of the ions comprising the surface molecules such as metal oxides and silver halides

Each of these categories is rich with examples of how the surface charge arises and changes with solution conditions such as:

- pH
- Ionic strength
- Addition of adsorbed ions
- Addition of surface-active agents (ionic or nonionic)
- Addition of reagents that chemically bond to the surface
- Order of addition of various species

The surface charge, more specifically the surface charge density, plays an active role in colloidal stability.

A simple definition of a colloid is this: A particle or molecule with at least one dimension less than, approximately, 1 μ and all dimensions greater than, approximately, 1 nm. Examples include most polymers, proteins, biomacromolecules, nanoparticles, latexes, carbon blacks, liposomes, exosomes, and many more. The list is too long to enumerate. An enormous number of commercial preparations fall into the category of colloidal dispersions.

The surface area per gram of material for colloidal-sized particles is orders of magnitude larger than it is for particles larger than, approximately, 1 μ. (See Chapter I, Figure I-1.) Therefore, surface effects, which are normally negligible, become dominant in the

---

[36] R. J. Hunter, "Zeta Potential in Colloid Science", Academic Press, 1981.

description of colloidal behavior. It is the surface charge density, for example, which is responsible for the repulsive force between charged, colloidal particles and therefore stability.

A colloidal dispersion is inherently unstable. On a macroscopic basis, it is predicted, from the second law of thermodynamics, that small particles will coalesce into big ones to reduce the surface area. On a microscopic basis, attractive forces are always present pulling the particles together. The attractive forces are present even for completely nonpolar particles. Such forces have their origin in the momentary, induced dipole effects that arise when nonpolar particles approach each other at random due to diffusion. These forces, variously called London, Van der Waals, or dispersion forces, are also responsible for the liquefaction of the noble gases such as helium, neon, and argon when the temperature is low enough.

The dispersion forces are often just as large as any permanent dipole forces of attraction. This may seem unusual until you look at some typical cases as enumerated in most physical chemistry textbooks. Another surprise, for the novice, is the long range over which these forces are important. For simple molecules, the dispersion forces, indeed, all the attractive forces vary as r^{-7} where r is the intermolecular distance. The attractive part of the potential energy varies as r^{-6}, but for colloidal particles, this simple description is insufficient. For colloidal particles the attractive force is the sum of the individual molecular forces, summed over all pairs that make up the particle. Although the result depends on shape, the key point is this: the potential energy of attraction varies between D^{-1} and D^{-2}, depending on shape, particle size dimensions, and distance of approach, where D is the shortest distance between particles. Thus, colloidal particles attract each other over much longer ranges than individual molecules, and this will lead to aggregation unless a repulsive force is present.

What is responsible for colloid stability? In general, there are two answers: electrostatic and steric repulsion. For suspensions in water and other highly polar liquids, electrostatic repulsion is the more common. The name "electrostatic" is misleading in the sense that the particles, solvent molecules, and ions are always in motion. It is used, however, because much of the theory is based on relatively simple electrostatic equations; and because it is assumed that the time averaged results can be calculated from an electrostatic theory. The result is known as the DLVO theory after the proposers--Derjaguin, Landau (Russia) and Vervey, Overbeek (The Netherlands).[37] Its success over the last 70 years is testimony to the reasonable character of the assumptions underlying the theory. Though few practical applications have been fully interpreted with the DLVO theory, it is the best foundation currently available for describing colloid stability, dispersion, and aggregation processes.

---

[37] Vervey, E. J. W. and Overbeek, J. Th. G., "Theory of the Stability of Lyophobic Colloids", Elsevier, Amsterdam, Netherlands (1948).

V.3 Electrophoresis

Due to the existence of surface charge, particles will move when placed in an electric field, E. This phenomenon is called electrophoresis. Microelectrophoresis is the common name given to the motion of charged colloidal particles in a liquid. Specifically, the particles move with an average drift velocity, V_e toward the electrode of opposite charge. At low field strengths the drift velocity is proportional to the electric field, and the proportionality is expressed in equation V-5 as $V_e = \mu_e \cdot E$, where μ_e is defined as the electrophoretic mobility.

There are three other related electrokinetic phenomena: electroosmosis, streaming potential, and sedimentation potential. Each has their own specific applications. For a colloidal dispersion the most useful technique is electrophoresis.

The electric field is commonly given in units of volts/cm (V/cm), and its value ranges from near 0 to a few tens of V/cm (except in nonpolar liquids where it can range up to a few hundred V/cm). The electrophoretic velocities that develop are in the range of 0 to a few hundred microns/second (μ/s). Mobility's are in the range of 0 to ± 10 Mobility Units.

Mobility units are commonly given in (μ/s)/(V/cm) or $\mu \cdot s^1 \cdot V^{-1} \cdot cm$. This mixture of length units--microns and centimeters--is awkward by SI standards, yet it results in convenient dimensions for mobility between 0 and ± 10. Conversions to CGS and MKS units are simple:

1 Mobility Unit (μ_e) = 1 $\mu \cdot s^1 \cdot V^{-1} \cdot cm$ = 10^{-4} $cm \cdot s^1 \cdot V^{-1} \cdot cm$ = 10^{-4} $cm^2 \cdot s^1 \cdot V^{-1}$ (CGS)

1 Mobility Unit (μ_e) = 10^{-4} $cm^2 \cdot s^1 \cdot V^{-1}$ = 10^{-8} $m^2 \cdot s^1 \cdot V^{-1}$ (MKS)

The MKS units are the preferred SI units. Nevertheless, the customary practice of using mobility units will be followed throughout the rest of this book as, indeed, this practice is followed in much, but not all, of the literature on this subject.

[Note: The Greek letter μ, pronounced mu, is sometimes used to designate both the electrophoretic mobility and a unit of length, namely, micron (10^{-6} meters). Here, the subscript e is used to designate the electrophoretic mobility, μ_e. In the scientific literature it is not always done this way. This presentation is consistent, but other works are not, and the reader should be careful.]

A particle with positive mobility means its surface (more specifically its shear plane) is positively charged; a negative mobility means the surface (shear plane) is negatively charged. Most surfaces are negatively charged under normal solution conditions, though by changing the solution conditions, a positive charge may result. The sign of the mobility, therefore, is a very important quantity. If a particle with a negative surface charge is mixed with a particle with a positive surface charge, the result is an unstable dispersion that will aggregate. (The later statement needs modification if long-chain, nonpolar polymers are also attached to the surface. These long chains provide steric repulsion, the name given to entropic and osmotic effects that drive the particles apart. In what follows, it will be assumed that only simple electrostatic repulsion is significant.)

When the mobility is zero, the electrostatic velocity is zero, and that implies that electrostatic repulsion is small if not zero. Under this circumstance particles will aggregate, because, again, the attractive forces are always present.

V.4 The Double Layer

In the early, simple theories of the charged solid-liquid interface, it was assumed that a layer of counter ions—oppositely charged ions to those on the surface—lined up to surround the surface. These counter ions were held rigidly in place by the electrostatic forces and moved with the particle. This model gave rise to the name electrical double layer. Figure V-4 shows a simple picture of this for a flat surface.

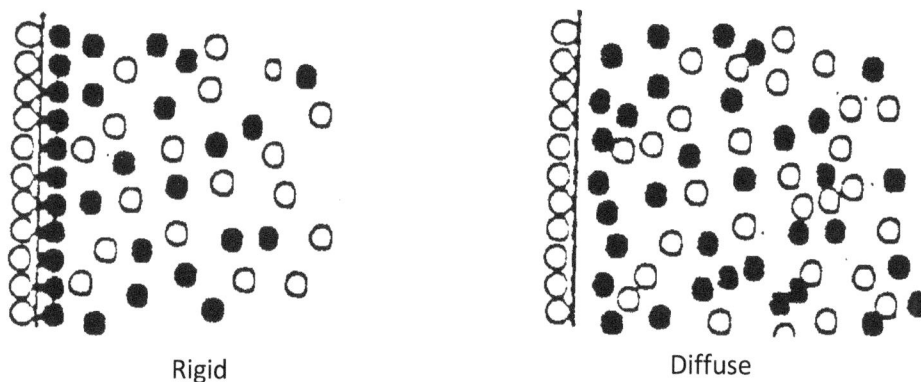

Rigid   Diffuse

Figure V-4: Rigid and Diffuse models of a simple Double Layer.

The name has been used ever since, although it was recognized that such a simple model could not suffice to describe all the known phenomena. For example, although the counter ions must be attracted to the surface more strongly than the co-ions—ions of the same charge as those on the surface—both the counter and co-ions in the bulk of the solution are continually jostled by the thermally driven, random motions of the solvent molecules. This picture leads to the model of the diffuse electrical double layer. The concentration of counter ions is higher near the surface, and it decreases steadily to the concentration in the bulk liquid. Likewise, the co-ions are depleted near the surface and their concentration increase until it reaches the level in the bulk liquid. Figure V-5 shows the concentration of co- and counter ions as a function of distance calculated from this simplest of models.

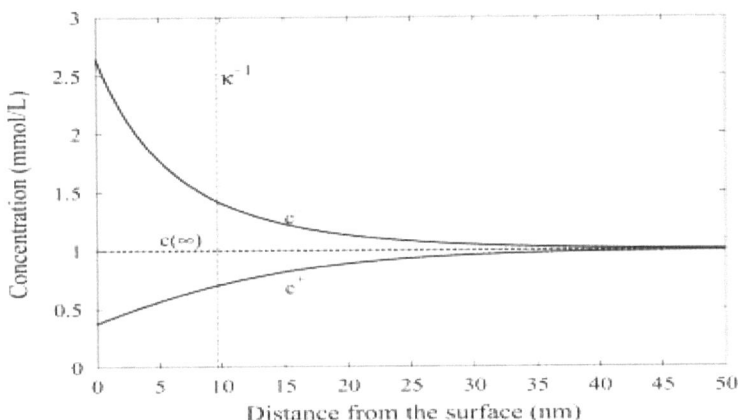

Figure V-5: Relative concentrations of co- and counter-ions near a surface.

Under a variety of simplifying, yet reasonable assumptions (see the references), this model leads to the following equation for the electrostatic potential Ψ as a function of distance x from the particle surface:

$$\Psi = \Psi_o e^{-\kappa x} \tag{V-8}$$

Where, Ψ_o is the potential at the surface and κ is a parameter that determines the decay of the potential with distance.

Although this equation is based on a much too simplistic model —many more sophisticated models are discussed in the book by Hunter (see his footnote #2)— it does embody several aspects of the diffuse electrical double layer theory that are important even in the more advanced theories. First, the zeta potential (discussed below) decreases in an exponential fashion like the electrostatic potential Ψ. Second, the parameter κ^{-1}, called the double layer thickness, plays a key role in determining the distance from the particle surface over which various electrical potentials are significant.

The double layer thickness is a function of the temperature, the dielectric constant of the liquid, and the ionic strength due to free ions (typically, added salt) in the bulk of the liquid. The double layer thickness is defined mathematically in Appendix Z2, and values are calculated under a variety of conditions. In Appendix Z1, the ionic strength is defined mathematically, and formulas are given for several types of electrolytes.

For 1:1 electrolytes such as KCl, it is useful to remember that the double layer thickness is approximately 100 nm at 10^{-5} Molar, 10 nm at 10^{-3} Molar, and 1 nm at 10^{-1} Molar. Thus, at low salt concentrations κ^{-1} is large, and the repulsive forces are long range. At high salt concentrations, κ^{-1} is small, and repulsive forces are short range. You can cause coagulation in an electrostatically stabilized colloid by adding enough salt.

The double layer thickness also plays a crucial role in more advanced theories where the calculation of the zeta potential depends on the value of the dimensionless product $\kappa \cdot r$, where "r" is the radius of the particle, later associated with r_H the hydrodynamic radius obtained from DLS and other techniques. This is discussed further in the section on zeta potential.

V.5 Electrokinetic Unit

Experimentally it is the velocity of the charged particle that is measured. The mobility is calculated from the measured velocity. Theoretically it is the zeta potential that is important in determining the magnitude of the electrostatic repulsive force. Unfortunately, there is no independent method for determining zeta potential; in microelectrophoresis it is always calculated from the mobility. In addition, there are no primary standards (only reference standards) available for determining the accuracy of a zeta potential calculated from a measured mobility. These two facts have profound consequences for the science of colloids, yet it does not hinder the usefulness of such measurements for many, many practical purposes associated with the determination of stability.

The zeta potential is the electrostatic potential at the surface of shear. It is not the potential at the surface of the particle. The surface of shear extends out from the particle surface. This surface separates the electrokinetic unit from the bulk of the solution. It is the kinetic unit that moves. The kinetic unit consists of the particle, ions adsorbed onto the surface, counter ions contained within the surface of shear, plus solvent molecules strongly attached to the surface ions and counter ions in the double layer. In addition, if long-chain surfactants or polyelectrolytes are attached to the surface, then they, and any solvent molecules strongly associated with them, also constitute part of the electrokinetic unit. It is this kinetic unit that moves as a single entity through the surrounding liquid.

Figure V-6 shows a flat surface as the y-axis (particle to left of axis) with the liquid to the right. In this example, only relatively small ions and molecules are near or attached to the surface, no long-chain polymers. Notice the various mathematical planes that can be drawn to distinguish various layers of entities associated with the surface. The exact location of these various planes depends on the model chosen to describe the solid-liquid interface. And, of course, any model is only an approximation to reality. There are no physical planes, only narrow regions over which something of interest changes. [Note: In many cases, the Stern and Shear planes are nearly coincidental.]

There is a sharp increase in viscosity between the bulk liquid and the kinetic unit at the shear plane, also called the shear surface. And, in principle, the dry particle size is less than the size of the kinetic unit. In practice, however, the difference is usually smaller than the random error in the measurement of the size, except for the case of small particles with long-chain surfactants or polymers attached.

[NOTE: The figure shows a uniform charge density, but that is often not the case. There are patches of charges and patches of nonpolar surface area in many cases. Even though particles are rotating rapidly, if the electrostatic repulsive force is not enough, these nonpolar patches may be sites for aggregation when particles collide.]

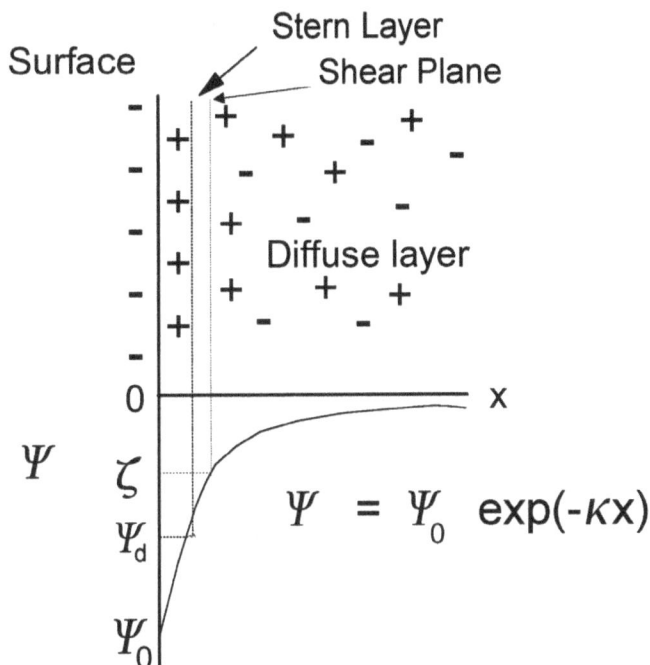

Figure V-6: Potentials in the Double layer model.

The extent of the inner layers is determined by the sum of ionic radii and the diameter of any strongly attached solvent molecules. This sum varies from 0.1 to 0.2 nm up to 1 nm for highly charged surfaces with large counter ions attached, assuming, again long-chain polymers are not attached to the particle. For dry particle sizes greater than 10 nm, the difference is usually less than the experimental error associated with the size measurement. Yet, it is the nature of the surface and the resultant inner layers that controls the stability of colloids like nanoparticles and proteins.

The size of the kinetic unit, also called the hydrodynamic size, is also the size that is measured in sedimentation and dynamic light scattering (DLS) measurements. In DLS measurements the diffusion coefficient of the kinetic unit is measured from which a hydrodynamic diameter is calculated. The hydrodynamic diameter is another name for the diameter of the electrokinetic unit.

(In the case of particles with long-chain polymers adsorbed onto the surface, the extent of the inner layer compared to the dry particle size may be significant. These are often called "Hairy" particles because the long-chain polymers, if they stick out into the inner layers, are said to resemble "hairs".)

V.6 Electrophoretic Mobility to Zeta Potential: Smoluchowski, Hückel and Henry Law Calculations

The force of repulsion, a measure of dispersion stability, is proportional to the square of the zeta potential. Zeta potential is calculated from mobility that, in turn, is calculated from electrostatic velocity. The basic equation for converting the mobility to zeta potential is given by:

$$BasicEquation \quad \mu_e = \frac{2}{3} \cdot \frac{\varepsilon_l}{\eta_l} \cdot \zeta \cdot f(\kappa \cdot r_H, \zeta) \tag{V-9}$$

where,

μ_e	=	Electrophoretic mobility, determined from measurement
ε_l	=	Liquid permittivity = $\varepsilon_o \cdot \varepsilon_r$ = 8.8542E-12 F/m $\cdot \varepsilon_r$
ε_o	=	Permittivity of free space
ε_r	=	Relative permittivity of liquid, also known as dielectric constant
η_l	=	Liquid viscosity
ζ	=	Zeta potential
$f(\kappa \cdot r_H, \zeta)$	=	Correction function dependent on $\kappa \cdot r_H$, ζ and particle shape
$f_l(\kappa \cdot r_H)$	=	Limit of f($\kappa \cdot r_H$, ζ) in several important cases, independent of ζ
κ^{-1}	=	Debye Length, also known as double layer thickness
r_H	=	Average hydrodynamic radius of assumed spherical particle

V.6.1 Limiting Cases where $f(\kappa \cdot r_H, \zeta) = f_l(\kappa \cdot r_H)$: Smoluchowski, Hückel, Henry

A. Smoluchowski Limit: $\kappa \cdot r_H \gg 1$ (large sizes, modest to high salt concentration)

$$\mu_e = \frac{\varepsilon_l}{\eta_l} \cdot \zeta \qquad , \quad f_l(\kappa \cdot r_H) = 3/2 \tag{V-10}$$

Example: Mobility is 1.00 μ·cm/V·s, in water at 25 °C. What is zeta potential in mV?

$$\zeta(mV) = 1,129.4 \cdot \frac{\mu_e(\mu \cdot cm/V \cdot s) \cdot \eta_l(cP)}{\varepsilon_r} \tag{V-11}$$

$$\zeta(mV) = 1,129.4 \cdot \frac{1.00 \cdot 0.8904}{78.54} = 12.8 \text{ mV}$$

B. Hückel Limit: $\kappa \cdot r_H < 1$ (small sizes, modest to low salt concentration)

$$\mu_e = 2/3 \cdot \varepsilon_l / \eta_l \cdot \zeta \qquad , \quad f_l(\kappa \cdot r_H) = 1 \tag{V-12}$$

Example: Mobility is 1.00 μ·cm/V·s, in water at 25 °C. The zeta potential in mV is given by:

$$\zeta(mV) = 1{,}694.1 \cdot \frac{\mu_e(\mu \cdot cm/V \cdot s) \cdot \eta_l(cP)}{\varepsilon_r} \tag{V-13}$$

$$\zeta(mV) = 1{,}694.1 \cdot \frac{1.00 \cdot 0.8904}{78.54} = 19.2 \ mV$$

This is 3/2 times as large as the value calculated using the Smoluchowski equation V-10.

C. Henry Equation: Spheres in 1:1 electrolyte and $\zeta \leq |25.7|$ mV[38]

$$\mu_e = \frac{2}{3} \cdot \frac{\varepsilon_l}{\eta_l} \cdot \zeta \cdot f_1(\kappa \cdot r_H) \tag{V-14}$$

Where, for $\kappa \cdot r_H > 1$, (moving towards the Smoluchowski Limit):

$$f_1(\kappa \cdot r_H) = 1 + \frac{1/2}{(1+2.5/\kappa \cdot r_H)^3} \tag{V-15}[39]$$

[NOTE: Errors in Hunter's 1981 book and repeated in Hiemenz & Rajagopolan book, for case of $\kappa \cdot r_H > 1$. The equation given in their books only holds for $\kappa \cdot r_H \gg 1$. But the above equation works well for $\kappa \cdot r_H > 1$.]

And for $\kappa \cdot rH < 1$, (moving towards the Hückel Limit):

$$f_1(\kappa \cdot r_H) = 1 + \frac{(\kappa \cdot r_H)^2}{16} - \frac{5(\kappa \cdot r_H)^3}{48} - \frac{(\kappa \cdot r_H)^4}{96} + \frac{(\kappa \cdot r_H)^5}{96} + smaller \ terms \tag{V-16}$$

[This equation is correctly given in books by Hunter and the 3rd edition of Hiemenz & Rajagopolan.]

V.6.2 Sample Calculations

A. Calculation of Ionic Strength, I:

You need this in the calculation of κ for use in $\kappa \cdot r_H$. The definition is:

$$I = \frac{1}{2} \cdot \Sigma c_i \cdot z_i^2 \tag{V-17}$$

Where c is the ion concentration in mol/L (molarity, M), and z is the valence of the ion: +1 Na^{+1}, +1 K^{+1}; +2 Ca^{+2}, +2 Mg^{+2}; +3 Al^{+3}; -1 Cl^{-1}, -1 $(NO_3)^{-1}$; -2 $(SO_4)^{-2}$; -3 $(PO_4)^{-3}$. The

---

[38] See Appendix Z3.

[39] http://www.kirbyresearch.com/index.cfm/wrap/textbook/microfluidicsnanofluidicsse87.html

ionic strength has the same units as concentration (molarity).

See Appendix Z1 for calculation of various ionic strengths.

Many solutions are simple 1:1 electrolytes like NaCl, where a 1 mM concentration yields a 1 mM ionic strength, but watch out for $Al_2(SO_4)_3$, a 3:2 electrolyte, because then a 1 mM solution has a 15 mM ionic strength.

B. Calculation of Kappa, κ:

$$\kappa = \left\{ \frac{2 \cdot e^2 \cdot 1000 \cdot I \cdot N_{avo}}{\varepsilon_l \cdot k_B \cdot T} \right\}^{1/2} \qquad \text{(V-18)}$$

Where,

e	=	Electronic Charge in Coulombs	=	1.6022×10^{-19} C
ε_l	=	Liquid permittivity, $\varepsilon_o \cdot \varepsilon_r$	=	8.8542×10^{-12} F/m·ε_r
k_B	=	Boltzmann's Constant	=	1.3807×10^{-23} J·°K^{-1}
T	=	Temperature in Kelvin (°K)		
I	=	Ionic Strength in units of mol/dm3 (mol/L or M, molar)		
N_{avo}	=	Avogadro's Number	=	$6.0221 \times 10^{+23}$ mol^{-1}

The number 1,000 appears so κ^{-1} will have units of nm.

See Appendix Z2 for more information on the calculation of κ. Substitution of the four constants into equation V-18 yields the following:

$$\kappa(nm-1) = 502.8 \times \left[\frac{I(mol/L)}{\varepsilon_r \cdot T(°K)} \right]^{1/2} \qquad \text{(V-19)}$$

C. Calculation of $\kappa \cdot r_H$, f_1, and which equation to use to convert mobility to zeta potential:

1 mM KCl in water 25 °C: So, $I = 0.001$ mol/liter, $\varepsilon_r = 78.54$, and $T = 298.15$ °K

$$\kappa(nm^{-1}) = 502.8 \times \left[\frac{0.001}{78.54 \cdot 298.15} \right]^{1/2} = 0.104 \text{ nm}^{-1} \text{ using equation V-19.}$$

r_H (nm)	1	10	100
$\kappa \cdot r_H$	0.104	1.04	10.4
Equation	Hückel	Hückel or Henry	Henry
f_1	1.000	1.000 or 1.013	1.262

Table V-1:
Calculating f_1
given $\kappa \cdot r_H$.

The value of f_1 for 20 nm diameter (10 nm radius) particle is 1.013 using Henry equation, but with a 1.3% error, it is 1.000 using the simpler Hückel equation. Remember, the Henry equation applies when $\zeta \leq |25.7|$ mV. (See Appendix Z3).

10 mM KNO_3 in water at 75 °C: So, $I = 0.010$ M/L, $\varepsilon_r = 60.97$, T = 348.15 °K

$$\kappa(nm^{-1}) = 502.8 \times \left[\frac{0.010}{60.97 \cdot 348.15}\right]^{1/2} = 0.349 \; nm^{-1} \text{ using equation V-19.}$$

Where the temperature variation of the dielectric constant for water was calculated using

$$\varepsilon_r = 78.54 \cdot 10^{-0.0022 \cdot (t-25)}$$
.

r_H (nm)	1	10	100
$\kappa \cdot r_H$	0.349	3.49	34.9
Equation	Hückel or Henry	Henry	Henry or Smoluchowski
f_1	1.000 or 1.003	1.099	1.41 or 3/2

Table V-2: Calculating f_1 given $\kappa \cdot r_H$.

The value of f_1 for 200 nm diameter (100 nm radius) particle is 1.41 using Henry equation, but with a 6.4% error, it is 1.5 using the simpler Smoluchowski equation. Remember, the Henry equation applies when $\zeta \leq |25.7|$ mV. (See Appendix Z3).

Globular protein in saline: So, $I = 0.154$ molar, $\varepsilon_r = 78.54$, T = 298.15 °K

$$\kappa(nm^{-1}) = 502.8 \times \left[\frac{0.154}{78.54 \cdot 298.15}\right]^{1/2} = 1.287 \; nm^{-1} \text{ using equation V-19.}$$

Consider a monodisperse, globular protein with $r_H = 7.5$ nm; thus, $\kappa \cdot r_H = 9.653$. Using equation V-15, the Henry equation result is $f_1(9.65) = 1.25$.

NOTE: While many people think for small particles one should use the Hückel equation, when the salt concentration is high, the Henry equation is more appropriate.

REMEMBER: The Hückel equation is equivalent to $f_1(\kappa \cdot r_H) = 1$ and the Smoluchowski equation is equivalent to $f_1(\kappa \cdot r_H) = 1.5$. For 1:1 electrolytes and $\zeta \leq |25.7|$ mV, we have the Henry equation as an alternative. The monodisperse, globular protein example above results in a value of f_1 half way between Hückel and Smoluchowski limits. Using the Hückel or Smoluchowski limits would result in either a +20% or -20% error, respectively.

V.7 Zeta Potential in Nonpolar Solvents or In Low Conductivity

In nonpolar solvents or when conductivity is low in water, what to do?

For nonpolar solvents, the conductivity is close to zero as is the ionic strength, κ, and $\kappa \cdot r_H$. Therefore, always use Hückel in nonpolar solvents.

How about water and a low conductivity (low salt concentration if any)? Assume a 1:1 electrolyte, and the conductivity is 10 µS/cm. What is κ? The conductivity of a 1 mM KCl solution is 147 µS/cm. Assume linearity below this concentration. Therefore, at 10 µS/cm the concentration is 0.068 mM = 0.000068 Molar.

$$\kappa(nm^{-1}) = 502.8 \text{ x } \left[\frac{0.000068}{78.54 \cdot 298.15}\right]^{1/2} = 0.0271 \text{ nm}^{-1} \quad \text{using Eq.(V-19).}$$

Use the Hückel equation, if r_H is < 36.9 nm (thus, $\kappa \cdot r_H$ < 0.999), or use the Henry equation, if $\kappa \cdot r_H$ > 1 and ζ < |25.7| mV.

Compare zeta potential calculated in a low dielectric constant, nonpolar liquid with that in water assuming they both had the same, low mobility of 0.100 mob units. Let the nonpolar liquid be dodecane at 25 °C, so the viscosity is 1.35 cP and the relative permittivity is 2.02. Using the Hückel approximation, equation V-13, the zeta potential is calculated as follows:

$$\zeta(mV) = 1,694.1 \cdot \frac{\mu_e(\mu \cdot cm/V \cdot s) \cdot \eta_l(cP)}{\varepsilon_r} = 1,694.1 \cdot \frac{0.100 \cdot 1.35}{2.02} = 114 \, mV$$

Whereas, assuming the same mobility in water, the result is calculated as follows:

$$\zeta(mV) = 1,694.1 \cdot \frac{\mu_e(\mu \cdot cm/V \cdot s) \cdot \eta_l(cP)}{\varepsilon_r} = 1,694.1 \cdot \frac{0.100 \cdot 0.8904}{78.54} = 1.92 \, mV$$

A similar large difference would be seen if the Smoluchowski or the Henry equation had been appropriate. The lesson is this: Whereas a low mobility results in a low zeta potential in water, in a nonpolar liquid the result may be a high zeta potential.

Low Mobility Can Still Lead to High Zeta Potential

Often electrophoretic mobility determinations are made in water. This translates to 12.8 mV per mobility unit <u>in water</u> using the Smoluchowski equation. [See equations V-10 and V-11.] Occasionally measurements are made in low dielectric constant, nonpolar solvents, or in viscous liquids either of which leads to a low mobility. But, as above, when calculating zeta potential, it may be surprisingly high.

Consider the following Table V-3 to get an idea of the effect on zeta potentials calculated in various liquids with a variety of viscosities and dielectric constants (relative permittivity)

Table V-3: Mobility ratio for particles with the same zeta potential in various media

Liquid	Viscosity	Dielectric Constant	Mobility Ratio
Water	0.89	78.54	1.00
Methanol	0.54	33	0.7
Glycerol	1.2	43	0.4
Toluene	0.56	2.38	0.05
n-Octane	0.54	1.96	0.04
Ethylene Glycol	17	40	0.03
1,4 Dioxane	1.26	2.24	0.02
Oleic Acid	26	2.46	0.001

The mobility ratio calculated here is just the dielectric constant of the liquid divided by its viscosity all normalized to the ratio for water (set to unity). For either high viscosities or low dielectric constant, this ratio decreases. This means, for the same zeta potential, the electrophoretic mobility decreases in proportion to this ratio. As the ratio decreases, it becomes harder and harder to measure the Doppler shift frequency (ELS). Just increasing the electric field strength is not the answer, because that causes heating and other problems. Using a new detection scheme (PALS) is the answer.

V.8 Optical mixing (Heterodyning) and Phase Analysis Light Scattering (PALS)

When using ELS, the Doppler shift doesn't produce a complete cycle, and cannot be determined, until the particle has moved at least a distance $> q^{-1}$. (Remember, $\mathbf{q} \cdot \mathbf{r}$ appears in the scattered electric field.) In water, at a small scattering angle to minimize diffusion broadening using a 640 nm laser, $q^{-1} = [640 \text{ nm}/4 \cdot \pi \cdot 1.33 \cdot \sin(15/2)] = 293$ nm. Thus, to use ELS a particle must travel under the influence of an electric filed at least that distance. For cases of low mobility that can take a lot of time or a large electric field must be applied. The longer a field is applied, the more electrode polarization can occur, changing the field the charged particle experiences. High electric fields (except in the case of nonpolar liquids or very low conductivity) can cause chemical changes and heating. Thus, a more sensitive technique is needed.

The phase of the scattered light, $\phi = \omega \cdot t$, changes by 2π, its maximum, when the particles moves a distance equal to its own diameter. If phase changes can be measured this provides a very sensitive technique since particles only have to move a fraction of their own size. Smaller fields and shorter times are possible.

From the relationship between phase and frequency, it follows that $d\phi/dt = \omega$, which, when applied to the Doppler shift frequency ωs, results in the following:

$$\frac{d\phi_s}{dt} = \omega_s = \boldsymbol{q} \cdot (\boldsymbol{V_e} \pm \boldsymbol{V_c}) \tag{V-20}$$

Where equation V-6 has been used with a wrinkle. If there is a small, field-independent collective velocity, $\boldsymbol{V_c}$, probably thermally driven, it too will result in a Doppler shift. When doing ELS, such velocities are also present, but usually too small to worry about. They result in the shifted peak moving a little more-or-less from its position due only to electrophoretic light scattering. The effect can be averaged out. Here, given the sensitivity of phase analysis, it is worthwhile to see how it affects phase analysis.

Using equation V-5 to substitute for $\boldsymbol{V_e}$, one finds:

$$\frac{d\phi_s(t)}{dt} = \omega_s = \boldsymbol{q} \cdot (\mu_e \boldsymbol{E}(t) \pm \boldsymbol{V_c}) \tag{V-21}$$

Where a time dependence of the phase has been added because the applied field will not be constant for PALS. A sinusoidal electric field with frequency ω_E is applied (note: other oscillating fields may be used), and after optically mixing the scattered light with a reference beam with frequency ω_o and phase ϕ_o, and integrating over the time t_E the field is applied, the result is the following[39,40]:

$$\Delta\phi_s = \langle A \rangle \boldsymbol{q} \cdot \boldsymbol{E_o} \langle \boldsymbol{\mu_e} \rangle \frac{cos(\omega_E t_E)}{\omega_E} \pm V_c t_E \tag{V-22}$$

The amplitude weighted, phase difference is plotted against the time the field is applied, t_E. The measurement also determines the average amplitude $<A>$. If there are no thermally driven velocities, V_c is zero, the plot is a cosine with the difference between the max and min related to the average mobility $<\mu e>$. Thus, only an average mobility can be determined. If there were a mixture of mobilities, ELS might be able to separate the shifted frequency peaks but PALS can only produce an average. Fortunately, except for mixtures, an average

---

[39] Tscharnuter, W. W., Applied Optics, **40**, 3995-4003, (2001).
[40] Miller, J. F., Schätzel, K. and Vincent, B., "The determination of very small electrophoretic mobilities in polar and nonpolar colloidal dispersions using phase analysis light scattering.", **143**, 532-554, (1991).

mobility is enough for most purposes.

Examples of PALS measurements are shown in Figure V-7 and V-8.

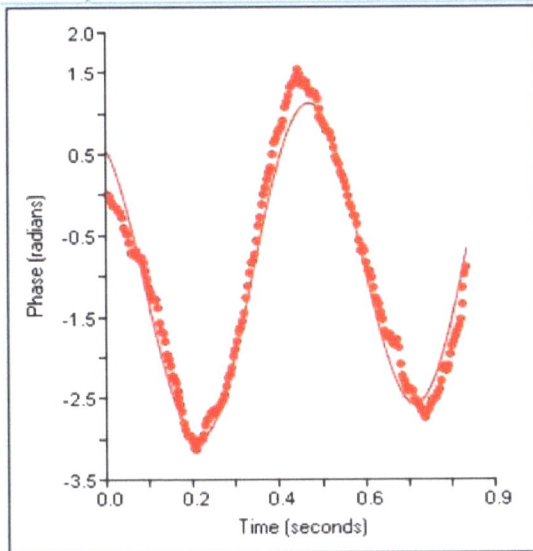

Figure V-7: PALS result on α-crystallin (a protein) from a cow eye. The suspending media is a high salt concentration of phosphate. The measured red data values are fitted to the solid curve. It is a cosine riding on a slightly increasing straight line accounting for V_c. The initial negative slope results in a negative, average mobility and zeta potential.

The y-axis is the difference in phase $\Delta\phi_s$; the x-axis is the time the field is on in one direction t_E.

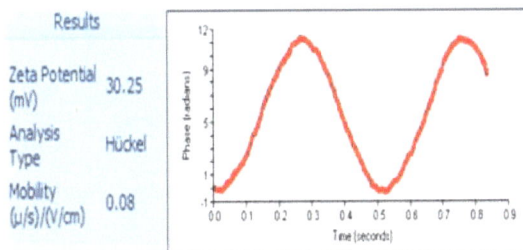

Results	
Zeta Potential (mV)	30.25
Analysis Type	Hückel
Mobility (μ/s)/(V/cm)	0.08

Figure V-8: PALS measurement of TiO_2 in toluene ($\varepsilon_r = 2.38$, $\eta = 0.558$ cP). The initial positive slope results in a positive, average mobility and zeta potential.

An example of the extreme sensitivity of the PALS technique is given in Miller, et. al.[40]. Results for multiple measurements on silica particles coated with poly(dimethyl aminoethyl methacrylate-co-poly(methyl methacrylate) in 1,4, dioxane ($\varepsilon_r = 2.24$, $\eta = 1.26$ cP) are given. The result is as follows: $<\mu_e> = (4.89 \pm 0.6) \times 10^{-11}$ m²/V·s = $(4.89 \pm 0.6) \times 10^{-3}$ mobility units. The random error is high, over 10%. And yet the mobility is about 0.005 mobility units, about 1,000 times smaller than the average mobility for particles in water. Such a small value could never be determined by a Doppler shift measurement using ELS.

PALS can be used for any mobility measurement but is especially useful when the mobility is low. If ELS shows more than one peak, PALS is not recommended unless an average value is enough, which is the common situation.

V.9 Electrophoretic Mobility to Zeta Potential: Advanced Treatments

All advanced treatments for spheres, like the Henry equation, require $\kappa \cdot r_H$. The implication is far reaching for a broad size distribution. At the low end of the size range, where $\kappa \cdot r_H < 1$ might hold, the Hückel equation applies; whereas, at the upper end of the size range where $\kappa \cdot r_H \gg 1$ might hold, the Smoluchowski equation applies. Ideally, one wants to separate the distribution by size, determine electrophoretic mobility (hoping it too is not broad), and apply the correction factor $f(\kappa \cdot r_H, \zeta)$ for that size class. Finally, use the corrected values to build up an accurate zeta potential distribution. To date, there is no single instrument that does this. While a fractionation by size followed by a mobility determination for each size fraction would work, there are no reports of this. Perhaps it is just not necessary.

One reason is that absolute zeta potential is rarely required to solve practical problems in surface chemistry. Such practical problems involve asking whether the surface is positive or negative, how it might change over time or with the addition of acid, base, surfactant or dispersing agent, and salt. And, most importantly, is it a stable dispersion? Answers to these problems require monitoring the relative change in zeta potential. For this reason, except for academic interest, one rarely encounters an attempt to calculate zeta potential beyond the use of Hückel, Smoluchowski, or a simple Henry equation for 1:1 electrolytes when $\zeta < |25.7|$ mV.

Indeed, the application of zeta potential instrumentation beginning in the 1940's and 1950's was almost exclusively for larger colloidal sizes and using the Smoluchowski equation even though, as we saw in V.6.2.C.2, in 10 mM salt, $\kappa \cdot r_H$ is not large enough to apply Smoluchowski unless the sizes are in the hundreds of nanometers and salt concentration even lower. Yet, Smoluchowski was applied so often, it became the default.

Smoluchowski Equation: The Default for Zeta Potential

Clearly, reporting a ζ should be accompanied by solution conditions <u>and</u> equation used to convert μ_e to ζ. It is the square of ζ that is proportional to the electrostatic repulsion between particles and therefore related to dispersion stability. (See Appendix Z7). Historically, such measurements were applied to large metal oxide and halide (e.g. AgI) particles, and thus began the near exclusive use of the Smoluchowski equation. It remains true today that a reported zeta potential, if unlabeled, is most likely the result of applying the Smoluchowski equation, whether its use is fully justified or not. Even when labeled as calculated using the Smoluchowski equation, its use may not be fully justified.

V.10 Advanced Treatments: Relaxation & Retardation (Wiersema et al., O'Brien & White)

The applied electric field on the double layer ions causes the liquid to move in a direction opposite to that of the particles. This is just like the electro-osmotic effect on macroscopic surfaces that do not move. But here it has a retarding effect on the electrophoretic velocity. The Smoluchowski equation accounts for retardation. The Hückel equation also accounts for retardation but is only true for $\kappa \cdot r_H < 1$.

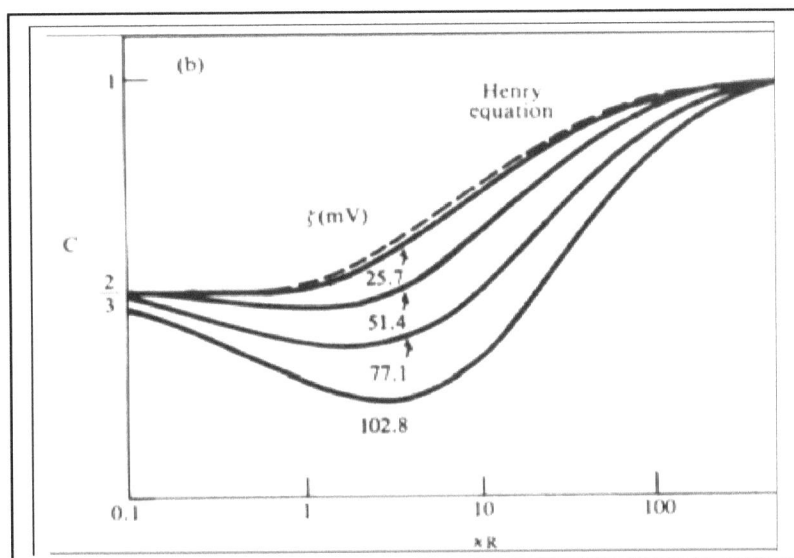

Figure V-9: $Y = C = f(\kappa \cdot r_H, \zeta)$ vs. $\kappa \cdot r_H$ for different $|\zeta|$ potential values according to Wiersema, et al. (1966) for a 1:1 electrolyte.

As the charge particle moves, it distorts the applied electric field. This gives rise to the relaxation effect. Both retardation and relaxation are part of the Wiersema, et. al.[41] and the O'Brien & White[42] treatments. Both these treatments for spheres are rather complete. The O'Brien & White one is particularly detailed but requires a computer program to apply it. Here, to get a flavor of the differences that can arise between the defaults usually used and something closer to the truth, consider Figure V-9.

The intercept on the left is the Hückel result, unity, which, as before holds for low $\kappa \cdot r_H$, independent of ζ. The intercept on the right is the Smoluchowski result, 3/2, which, as before holds for large $\kappa \cdot r_H$, independent of ζ. Notice further the Henry equation (dashed curve) applicable for 1:1 electrolytes and $\zeta < |25.7|$ mV. Notice how it is quite close to the more complete Wiersema solution for low ζ over the entire range of $\kappa \cdot r_H$. All three classical solutions

---

[41] Wiersema, P. H., Loeb, A. L. and Overbeek, J. Th. G., Journal of Colloid Interface Science, **22**, 78, (1966).

[42] O'Brien, R. W. and White, L., J. Chem. Soc. Faraday II, **26**, 1607-1626, (1978).

are in excellent agreement with the Wiersema solution.

Consider how different the resulting ζ potential values will be for a specific case: 1mM KNO_3, water, 25 °C, 280 nm average diameter of a broad size distribution, average mobility -3.46 mob units.

Use equation V-19 to determine κ (nm^{-1}) = 502.8 x $\left[\dfrac{0.001}{78.54 \cdot 298.15}\right]^{1/2}$ = 0.1039 nm^{-1}.

Therefore, $\kappa \cdot <r_H>$ = 0.1039 x 140 = 14.5 > 1. Thus, use equation V-15 to determine $f_1(\kappa \cdot r_H)$ = 1.311 for use in the Henry equation. Here are the results using different equations and interpolating from the graph in Figure V-9:

Equation	ζ (mV)	Comment
Smoluchowski (V-11)	-44.3	But $\kappa \cdot r_H$ not >> 1
Hückel (V-13)	-66.5	But $\kappa \cdot r_H$ not < 1
Henry (V-14)	-50.7	But it turns out, ζ not < |25.7| mV
Wiersema, Fig. V-9	-51.8	Interpolating f on ζ = |25.7| mV curve at $\kappa \cdot r_H$ = 14.5
Wiersema, Fig. V-9	-55.4	Interpolating f on ζ = |51.4| mV curve at $\kappa \cdot r_H$ = 14.5
Wiersema, Fig. V-9	-62.4	Interpolating f on ζ = |77.1| mV curve at $\kappa \cdot r_H$ = 14.5
Wiersema, Fig. V-9	-73.1	Interpolating f on ζ = |102.8| mV curve at $\kappa \cdot r_H$ = 14.5

Table V-4: Differences in ζ potential using different equations.

The classic, first three equations are not strictly applicable as noted in the comments. The next four are answers obtained assuming in turn each of the Wiersema solutions is correct, graphically interpolating the y-axis value of f corresponding to the intersection of each curve at $\kappa \cdot r_H$ = 14.5. Notice that the result at -55.4 mV is relatively closer to the curve labeled -51.4 mV (absolute value) than any of the other answers are to their corresponding curves. Thus, to 1st order ζ = - 55 mV.

A second order correction can be made. A plot of the four Wiersema values in column two of Table V-4 as the y-value against the corresponding set in column three yields a smooth curve which can be well fitted to a parabolic function:

$$Y = -0027 \cdot X^2 - 0.0695 \cdot X - 51.825$$

When Y = X, the consistent, 2nd order result is -56.5 mV. This is 27.5% more negative than the Smoluchowski result and shows once again that most reported zeta potentials are not absolute but relative.

If the particles are spheres, and narrowly distributed in size around r_H = 140 nm, it would be worthwhile trying to make such a correction. But for determination of relative changes it is not worth the effort. And in this example, the size distribution is not narrow. So, using the Smoluchowski result is acceptable to monitor relative changes.

V.11 Calculating the Charge on the Particle

Often one sees zeta potential instrumentation advertised as useful in determining the charge on the particle. This is not strictly true. The surface potential Ψ_0 (Fig. V-6) is related to the surface charge and the point of zero charge (PZC). But it is zeta potential that is related to charge at the shear plane and the isoelectric point (IEP), the solution conditions when $\zeta = 0$. [Because many metal oxides change zeta potential with pH, the IEP is often, but incorrectly, defined to be the pH at which $\zeta = 0$. More generally, the IEP is the concentration of the potential determining ion when $\zeta = 0$.

For example, $\zeta = 0$ when pBa7 for $BaSO_4$ and this has nothing to do with pH.]

To determine charge, in the simplest case, balance the electrical force, $f_{elec} = Q \cdot E$, against the frictional force of resistance $f_{fric} = 6\pi \cdot \eta \cdot r_H \cdot V_e$, where Q is the total particle charge (Z·e), E the applied electric field, η the liquid viscosity, r_H the hydrodynamic particle radius, V_e the electrophoretic velocity, and Z the valence or number of charges. Since electrophoretic mobility $\mu_e = V_e/E$, the valence or number of charges Z can be calculated from the following equation:

$$Z \cdot e = 6\pi \cdot \eta \cdot r_H \cdot \mu_E \qquad \text{(V-23)}$$

Accounting for relaxation and retardation as well as for a radius r_b for water molecules and buffer ions moving with the surface of the particle, Winzor[43] adds the usual corrections and finds the modified result:

$$Z^*_{eff} e = 6\pi \cdot \eta \cdot r_H \cdot \mu_E \cdot \frac{1 + \kappa(r_H + r_b)}{f(\kappa \cdot r_H) \cdot (1 + \kappa r_b)} \qquad \text{(V-24)}$$

Z^*_{eff}	=	Effective number of charge at shear plane surrounding particle
e	=	Charge on one electron, 1.602×10^{-19} Coulombs
η	=	Viscosity of liquid
r_H	=	Particle radius, if protein measured or estimated using $r_H = 0.031 \cdot M^{0.43}$
r_b	=	Average radius of ions surrounding particle, often estimated as 0.25 nm
μ_E	=	Measured electrophoretic mobility
κ	=	Inverse Debye Length, requires ionic strength, I
$f(\kappa \cdot r_H)$†	=	Henry's Function, true strictly only for 1:1 electrolytes and $\zeta \le |25.7 \text{ mV}|$

†This is to distinguish it from the complete correction factor $f(\kappa \cdot r_H, \zeta)$. Use Wiersema, Loeb & Overbeek (footnote 41) or O'Brien & White (footnote 42) to estimate a better correction factor.

---

[43] D. J. Winzor, Analytical Biochemistry, **325**, 1-20, (2004).

V.11.1 Sample calculations using equation V-24

I.   Calculate Z^*_{eff} for the protein lysozyme (r_H = 1.87 nm) at 25 °C in saline (154 mM NaCl & η = 0.890 cP). Assume the measured mobility is -2 μ·cm/V·s = -2 x 10⁻⁸ m²/V·s and there is no buffer layer adding to hydrodynamic size. Calculate κ using equation V-19.

$\kappa \cdot r_H$ = 1.29·1.87 = 2.41; $f(\kappa \cdot r_H)$ = 1.059 assuming Henry equation is applicable.

Now consider two cases:

Assume r_b = 0:

Z_{eff}^*e = 6·π·(0.000890 kg/m·s)·1.87 x 10⁻⁹ m·(-2 x 10⁻⁸ m²/V·s)·[(1+2.41)/1.059] = -2.019 x 10⁻¹⁸ kg·m²/V·s²

Moving 1 coulomb through 1 volt produces 1 Joule of work: C·V = 1 J(kg·m²/s²)

Therefore, Z_{eff}^*e = -2.019x10⁻¹⁸ Coulombs. The number of charges Z_{eff}^* = -2.019x10⁻¹⁸ Coulombs/1.602x10⁻¹⁹ Coulombs/charge = 13 negative charges (rounded to nearest whole integer) at the shear plane.

Assuming r_b = 0.25 nm:

1 + κ(r_H + r_b) = 1+ 1.29·(1.87 + 0.25) = 1 + 2.73 instead of 1 + 2.41 in numerator and

1 + $\kappa \cdot r_b$ = 1 + 0.323 instead of 1 in denominator

The result is:

Z_{eff}^*e = -1.671 x 10⁻¹⁸ kg·m²/V·s² . So Z_{eff}^* = -1.671 x 10⁻¹⁸/1.602 x 10⁻¹⁹ = 10 negative charges, rounded, at the shear plane.

Thus, you can't ignore r_b for proteins and other small particle sizes because it has a large impact on the calculated number of charges.

II.   Calculate Z^*_{eff} for a colloid reference material with r_H = 140 nm and negligible r_b, in 1 mM KNO_3 with μ_e = -3.46 mob units. As before, κ = 0.1039 nm⁻¹ so $\kappa \cdot r_H$ = 14.5 and f_1 = 1.311 if Henry applied and, using proportions, 1.176 along the ζ = 51.4 mV curve in Figure V-9.

Z_{eff}^*e = 6·π·(0.0008904 kg/m·s)·140 x 10⁻⁹ m·(-3.46 x 10⁻⁸ m²/V·s)·[(1+14.5)/1.176]

Z_{eff}^*e = -1.072 x 10⁻¹⁵ C, so Z_{eff}^* = -1.072 x 10⁻¹⁵/1.602 x 10⁻¹⁹ = 6,690 negative charges.

That is a lot of charges, but the particle size is large. So, what might the surface look like? Assuming charges were evenly spread over the surface, what would be the surface charge density, σ?

σ = 6,690/(4π·140²) = 0.0272 charges/nm² or every 36.8 nm² there is one charge. That is a square of about 6 nm x 6 nm or 60 Angstrom x 60 Angstrom. Now charges are part of ions associated with atoms (i.e. H^+) or small groups of atoms like a carboxyl group. The sizes of these entities are small, perhaps ionic radii of 0.1 to 0.3 nm. Therefore, most of the surface is not charged at all. This is one example of what is meant by the "charge—patch" model of the surface: Lots of nonpolar space where colliding particles can still aggregate.

Chapter VI: Surface Zeta Potential: SZP

VI.1 Introduction

Characterizing charged surfaces has many uses. In Chapter V, the large surface to volume ratio for submicron (colloidal) particles, a subset of which is nanoparticles, was shown to play an ever-increasing role in physical properties. The charged micro particle surface is often the key to understanding the stability of such dispersions with zeta potential a measure of the repulsive forces between such particles.

However, macroscopic surfaces in contact with a liquid can also develop a charge. Such surfaces can be plastic sheets, glass, ceramic, membrane, hair, fiber, even particles say tens of microns and larger, indeed anything else that is large compared to one micron. They develop charge, or it is added in the same way as it happens with submicron particles: covalently bonded acids, bases, and polar groups or added wetting and dispersing agents. In the presence of a liquid, most often water, such polar groups give rise to a Debye double layer and the resulting zeta potential associated with it. Indeed, the walls of charged sample cells of glass, quartz or plastic develop such charged surfaces, and this gives rise to electro-osmotic flow, an effect that must be either avoided or dealt with when making electrophoretic mobility measurements from which zeta potential is calculated.

So, characterizing the sign and magnitude of the charge effects on such macroscopic surfaces has real value. Some simple examples include toner in copy machines sticking to paper and dust and dirt on glass windows in buildings, motor vehicles, and residential dwellings. The first example is one of attraction: solid surface of the opposite sign as the particles. The second example is one of repulsion: solid surface of the same sign as the particles.

The classic method for determination of surface zeta potential is embodied in the streaming potential measurement. The simplest configuration consists of a large, flat cell, a sandwich really, with a very narrow middle section. As liquid at pressure is forced through the narrow channel, a voltage difference is registered at electrodes placed at either end of the cell. The voltage is plotted as a function of the pressure difference. And the slope is related to zeta potential. But there is another way to characterize the charge on macroscopic surfaces.

A macroscopic surface charge can be characterized using light scattering, probe particles, and a special version of the electrode assembly used in PALS. The latter will be explored first.

VI.2 Surface Zeta Potential (SZP) Electrode Assembly

A typical SZP electrode assembly, courtesy of Brookhaven Instruments, is shown in Figure VI-1. The sample is a relatively smooth solid, cut to 4.5 mm by 9 mm and up to 2 mm thick. It is attached (adhesive or glue) to a PEEK, (poly)etheretherketone, substrate that is screwed to a piston which, with precision gears, is lowered and raised between two palladium electrodes. When this assembly is inserted into a sample cell with liquid containing probe particles, the stage is set for an SZP measurement.

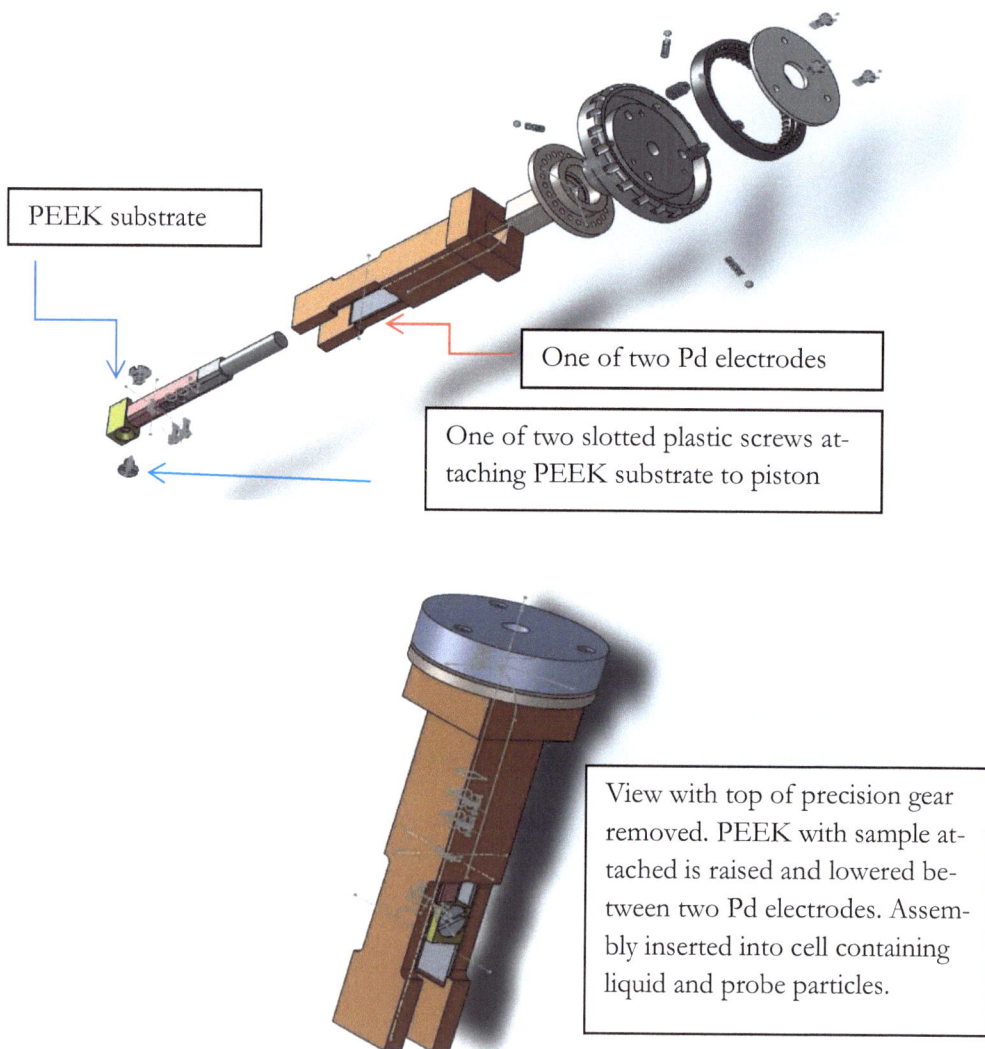

PEEK substrate

One of two Pd electrodes

One of two slotted plastic screws attaching PEEK substrate to piston

View with top of precision gear removed. PEEK with sample attached is raised and lowered between two Pd electrodes. Assembly inserted into cell containing liquid and probe particles.

Figure VI-1: SZP electrode assembly, courtesy of Brookhaven Instruments. Top picture is exploded view. Bottom is compact view. It is this part, once sample attached, that is lowered into sample cell.

A drawing of the raising and lowering of the surface of interest is shown in Figure VI-2. The sample is attached (glue or adhesive) to the substrate, which is in the liquid surrounded by the cell. The electric fields are shown as parallel arrows from the right Pd electrode (positive) to the left Pd electrode (negative). The focused laser beam is shown in the middle of the cell. Its diameter is approximately 100 micron.

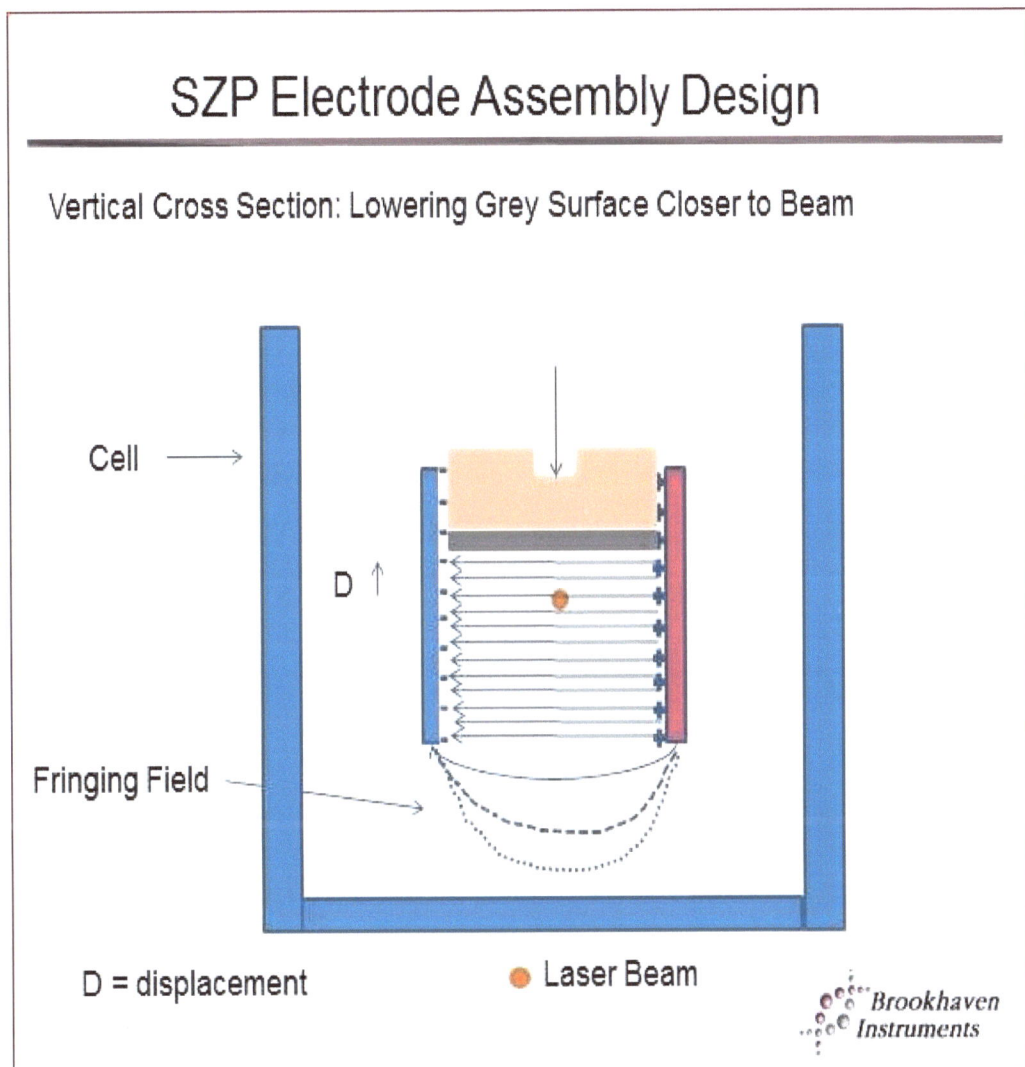

Figure VI-2: Relative position of sample surface (grey) as it is lowered and raised with respect to the beam.

The field lines are tangential to the surface of the sample of interest. Thus, any charge on the surface will generate electro-osmotic flow very close to the surface and thereby affect the electrophoretic movement of the particles close to the surface. Since the electrophoretic movement is determined by monitoring the scattering of probe particles in the laser beam, if that surface is close enough to the beam, the surface charge on the sample's surface will affect the electrophoretic flow of the probe particles and thus the electrophoretic zeta potential. Far from the surface, the effect is negligible, and the electrophoretic zeta potential should be that of the probe particles unhindered. Extrapolated to zero displacement, the resulting electrophoretic zeta potential of the probe particles, as perturbed by the surface electro-osmotic flow, is called the surface zeta potential, SZP

VI.3 Probe Particles

For surfaces that are negatively charged, use a negatively charged probe particle such as Spherotech's sulfonate stabilized polystyrene latex of 820 nm diameter. See www.spherotech.com. For surfaces that are positively charged, use a positively charged probe particle such as a common fabric softener. [The one from Seventh Generation, Inc., 60 Lake Street, Burlington, Vermont, USA works well. Other fabric softeners are also likely to contain positively charged particles.]

If you assumed the surface was negative but the SZP turns out positive, likely your assumption was wrong. And that means the negative particles, during the time they were in contact with the positive surface, were slowly neutralizing it. Thus, you are unlikely to get repeatable SZP measurements. Start over with a positively charged probe particle and a fresh surface. The same is true with a negative surface and positively charged probe particles.

VI.4 Surfaces

An ideal surface is a smooth, one-sided, adhesive tape. The adhesive side makes it very easy to attach to the substrate. It also makes it easy to remove and reuse the substrate. You can cut the tape to the exact dimensions required using a knife or scissors after it is applied to the substrate.

Double-sided sticky tape is useful with other samples; however, the sample itself will likely be cut to the correct size prior to pressing it onto the substrate. Use lab gloves to avoid adding any finger oils or other contaminants adhering to the surface.

WARNING

The sample surface area is not very large. Fingerprints, or any other contamination, may dramatically affect the results. Check that lab gloves do not carry any contamination.

The tiniest drop of silicone or epoxy glue can be used to secure sample to substrate. Obviously, make sure the sample is sized correctly before gluing. Make sure the sample is parallel to the substrate and not tilted because the glue has not spread evenly.

VI.5 Making Measurements

In principle, measurements can be made with ELS (Electrophoretic Light Scattering) or PALS (Phase Analysis Light Scattering). Those shown below were made with PALS.

The initial goal in such measurements is to find the zero position. This is defined as the position where the surface is at the middle of the laser beam. Starting above the beam, where the count rate is highest, establish an average value by measuring for many seconds. Lower the sample until the count rate is a minimum, indicating the beam is striking the sample or substrate. Note that it is only a few steps from highest to lowest count rate, where steps are nominally 20 to 25 micron depending on the electrode assembly used.

Figure VI-3 shows count rate vs. time as the sample is lowered and raised for a BI-SZP electrode assembly from Brookhaven Instruments. Despite the noisy signals, the zero is determined to within +/- 20-25 micron. This uncertainty adds little error to the result.

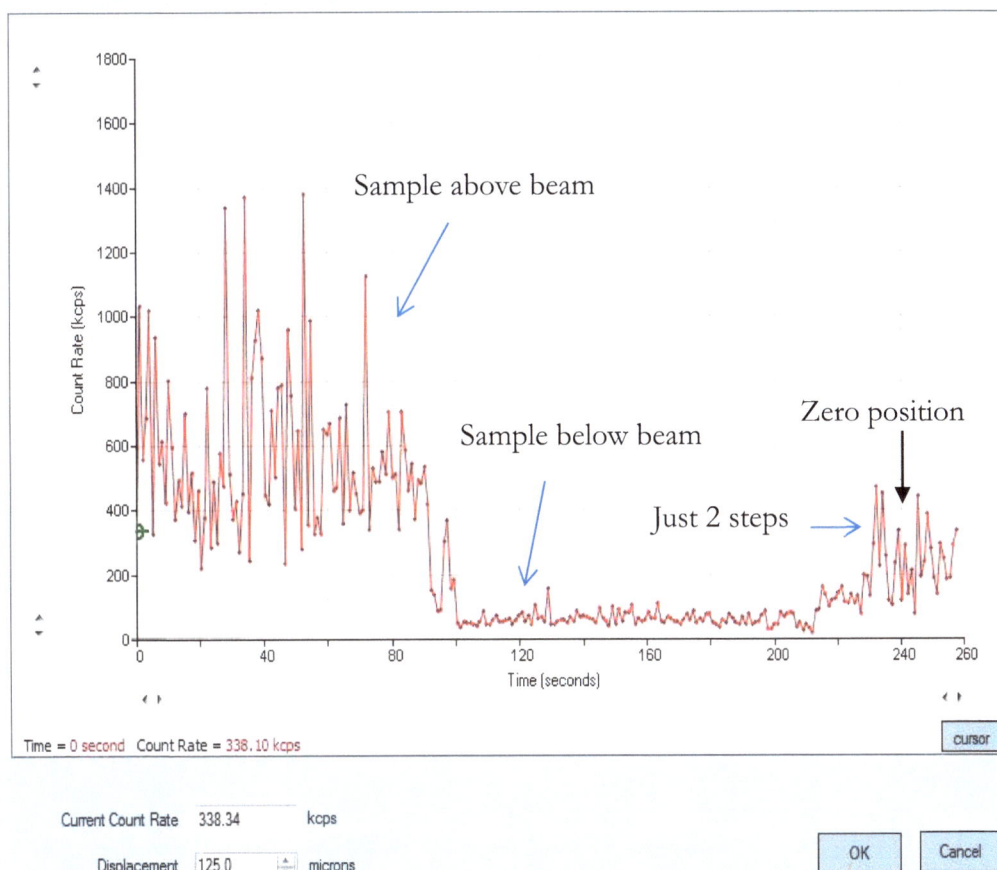

Figure VI-3: Finding the zero-displacement position.

Once you have found zero displacement, enter the first displacement, 100 μ or 125 μ, at which the measurement will take place. Make sure to raise the sample to that position before starting the measurement. At every displacement from zero, you can make one measurement; however, making three measurements helps to characterize the variations.

Continue making measurements at several displacements out to 500 μ to 1,000 μ. Figure VI-4 shows triplicate measurements taken at each of four displacements: 125, 250, 375 and 500 μ. The linear fit yields an intercept of -16 mV. This is the SZP, the value at zero displacement. The data are fit to the following equation: $\zeta = m \cdot D + SZP$.

Here ζ is the zeta potential determined from the measurement, m is the slope, D is the displacement, and SZP is the intercept at zero displacement.

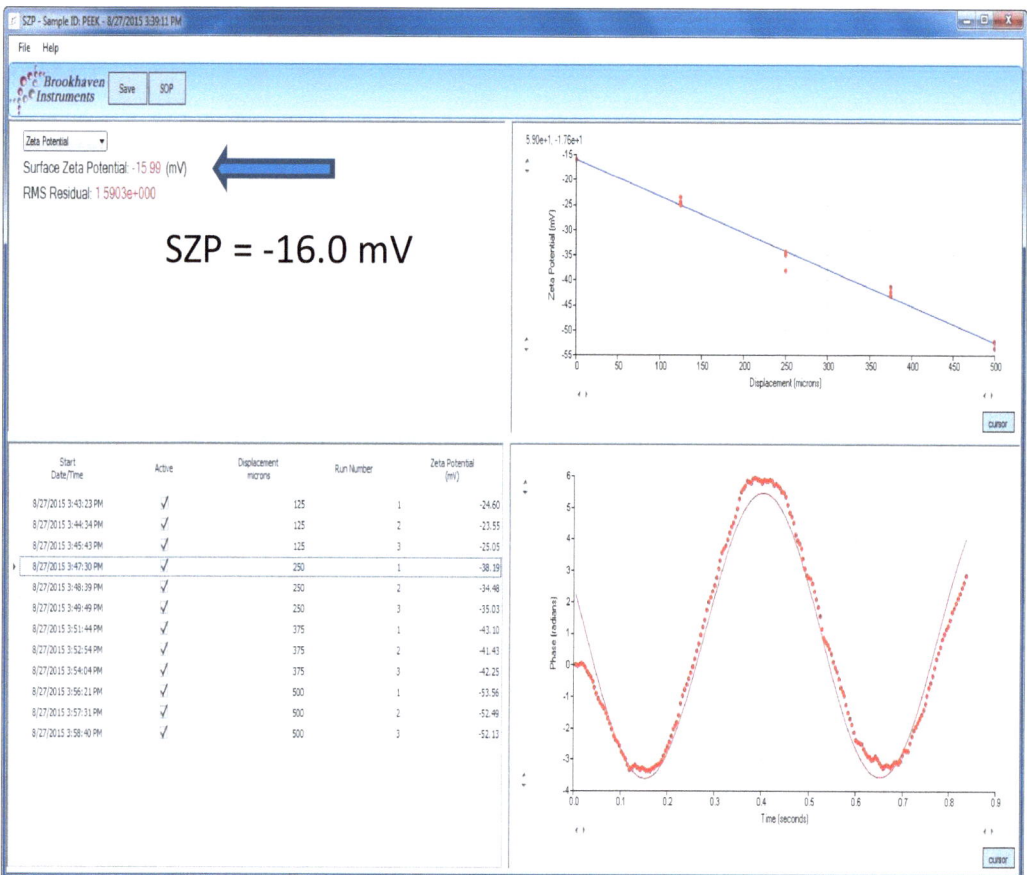

Figure VI-4: A surface zeta potential measurement out to 500 μ.

VI-6 Measurements at Large Displacements

What happens when you make measurements far from the laser beam? Here is an example:

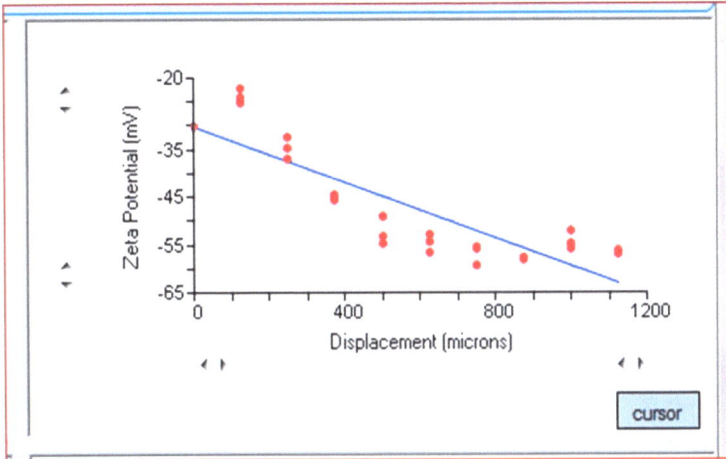

Figure VI-5:
A surface zeta potential measurement out to 1,200 micron.

The straight line fit to all the points is obviously not a good fit at all. The intercept around -30 mV is not a good representations of the SZP. Clearly, the last groups of data points, starting at 625 μ out to 1,125 μ, are not affected by the position of the sample. The sample is far enough away (large displacement) from the laser beam that the electrophoretic value of the probe particles is not affected by the sample surface. The right way to treat this data is to exclude all the far displacements from the fit. In this case, fit only up to 500μ. This is shown in Figure VI-6.

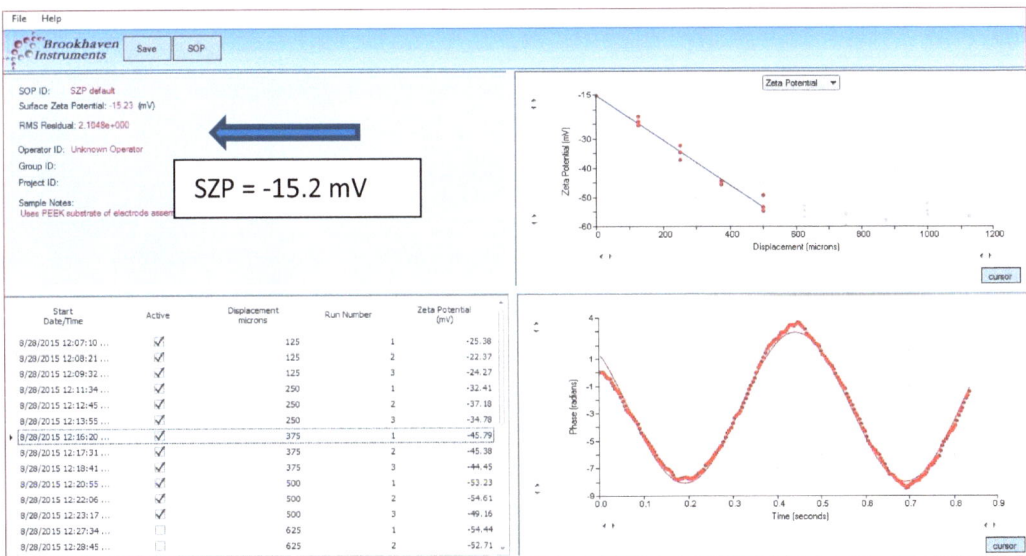

Figure VI-6: Ignore far displacement measurements.

The data points beyond 500 μ are essentially flat as shown in the enclosed box in the upper right graph. Using only the points at or below 500 μ, the fit is now much better. The SZP is about -15.2 mV.

The displacement beyond which the probe particles retain their unhindered electrophoretic mobility and zeta potential is probably dependent somewhat on the SZP itself. A surface with a lot of charge density and a large SZP will probably influence probe particles beyond 500 μ. To account for this, measure out to perhaps 1 mm but fit only the points that follow the initial straight-line drop.

VI-7 Using SZP to Compare Coated and Uncoated Surfaces

The surface of freshly uncovered (peel off the tape in the roll) 3M Frosted Scotch Magic™ Tape is relatively nonpolar. A drop of DI water balls up on such a surface indicative of a high contact angle (angle between solid surface and liquid) between the nonpolar surface and the polar liquid, water. If the surface was treated with a surfactant, the drop would have a lower contact angle and spread out over the surface. It would not look like a spherical ball. Figure VI-7 is a picture of such a surface with a droplet of DI water (on the right) next to a droplet of an aqueous solution of sodium dodecylbenzene sulfonate. The SDBS droplet is spreading.

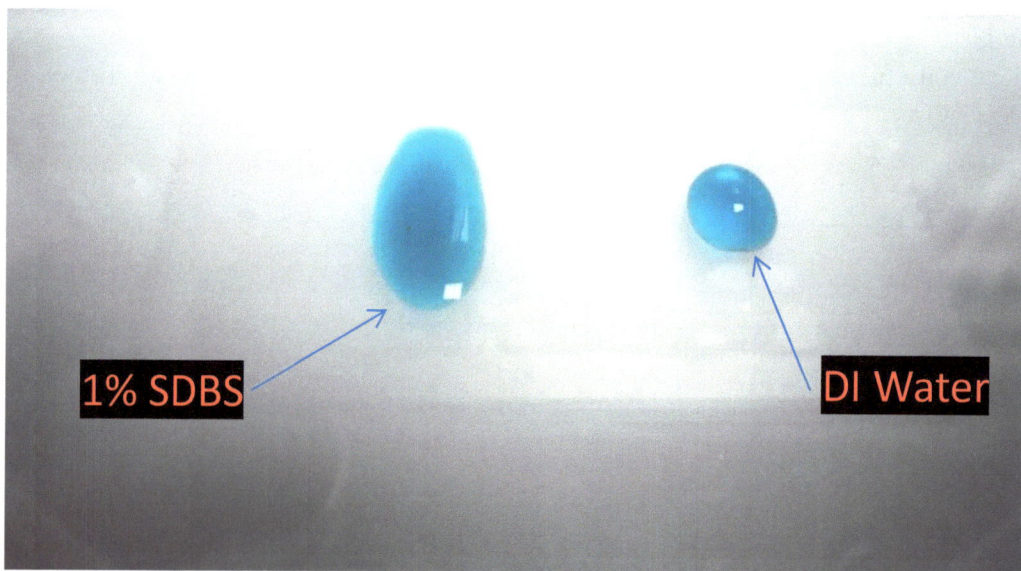

Figure VI-7: DI water droplet does not spread as much as a surfactant solution on a relatively nonpolar tape surface. Both droplets are blue-green colored for contrast.

Although not obvious at first, the droplets are lying flat, face up on a piece of the frosted tape with a white plastic background provided for enhanced contrast.

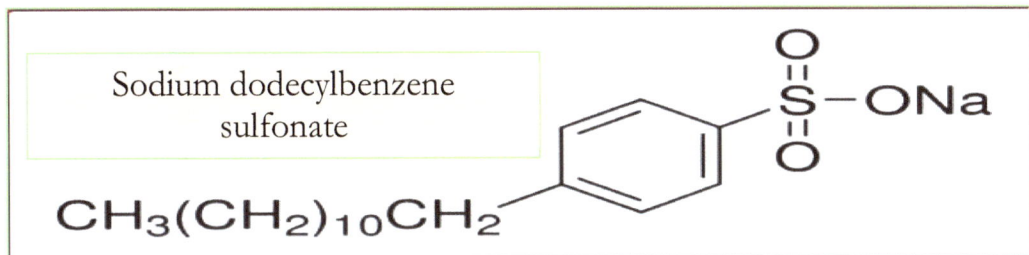

The nonpolar tail of SDBS physically adsorbs onto the nonpolar tape surface leaving the polar sulfonate end sticking up. When surrounded by water, the sodium ion will diffuse away leaving a negatively charged surface.

Figure VI-8 shows the SZP graph with no surfactant added to the tape surface. The SZP is close to zero.

Displacement (μ)

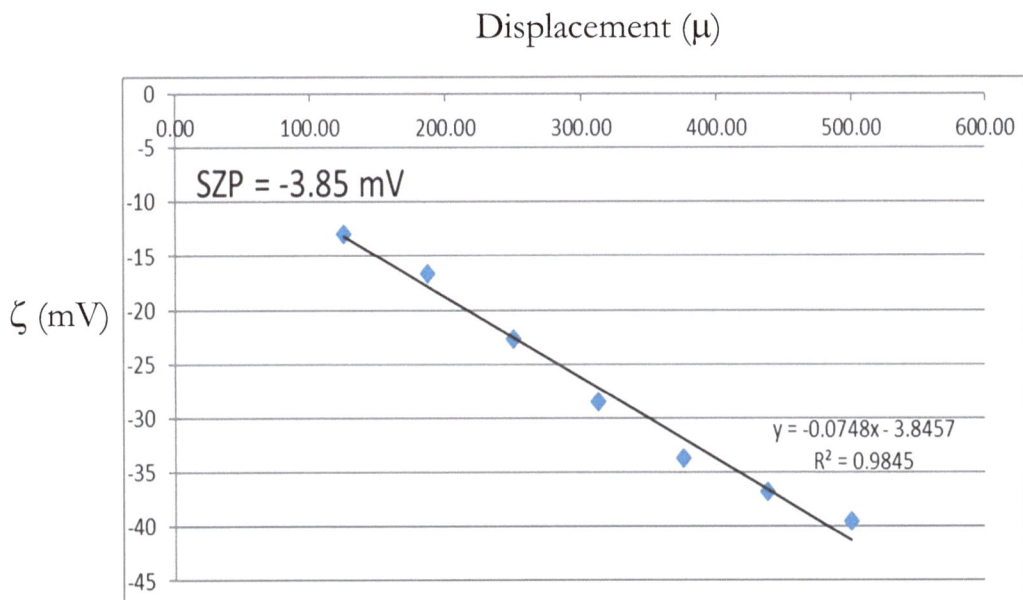

Figure VI-8: Zeta potential vs. Displacement of nonpolar sample surface.

Figure VI-9 shows the SZP graph after 1% SDBS was added to the nonpolar surface and allowed to spread for several minutes. The SZP is almost -19 mV.

Displacement (μ)

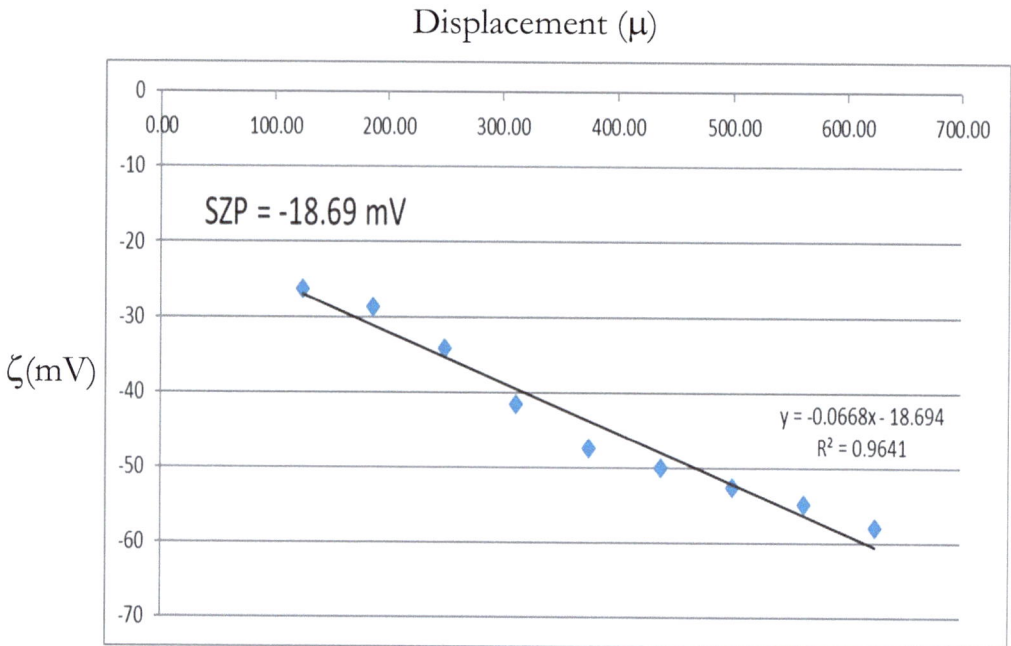

Figure VI-9: Zeta potential vs. Displacement of nonpolar sample surface after coated with 1% SDBS.

Clearly the surfactant has attached to the surface and has imparted it with a negative charge.

VI-8 The Meaning of SZP

What does surface zeta potential mean? When liquid encounters a macroscopic surface, if there are charges associated with that surface, then a double layer is formed just like it does around small particles. Thus, there is an electrostatic potential difference between the stationary plane near to the macroscopic surface and a point out in the liquid surrounded by free ions. This is the same zeta potential definition used with small particles.

A classic streaming potential device can be used with macroscopic surfaces. The measured streaming potential is converted into a zeta potential associated with the charge density on the surface. This is the most commonly thought of surface zeta potential.

With the SZP measurement, the experimental setup is different. Because of the tangential electrical fields cutting across the macroscopic surface, electroosmotic flow is induced near the surface due to charges on the surface. It is this flow that interacts with probe particle electrophoresis in an SZP measurement. The closer the probe particles examined are to the surface, the stronger the interaction; the further away the probe particles examined are from the surface, the weaker the interaction. And if the resulting zeta potential is extrapolated to zero displacement, meaning at the macroscopic surface, it is called the SZP.

Is it the same as the zeta potential calculated from a streaming potential measurement? Not enough work has been reported to know if it is. But the following measurements shown in Figure VI-10 indicate that at least the SZP is independent of the probe's zeta potential. Thus, the SZP is a characteristic of the macroscopic surface and not the probe.

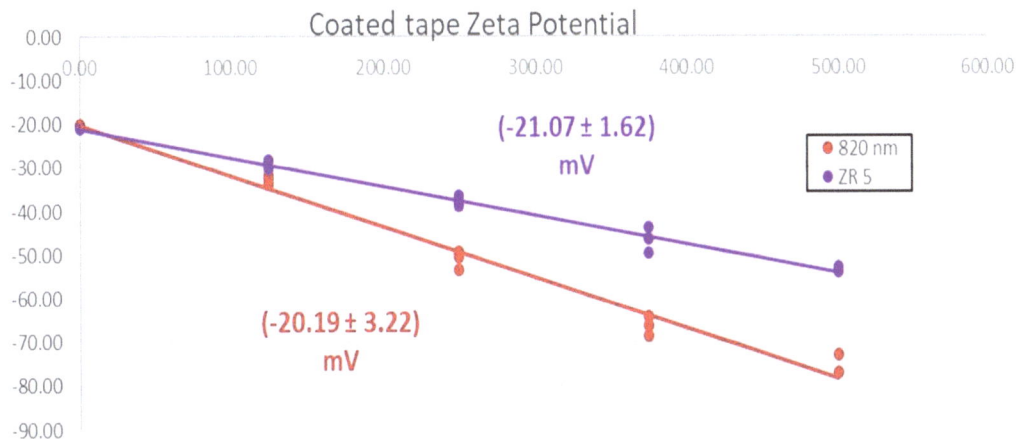

Figure VI-10: Zeta (mV) vs Displacement (μ) using different probes on same surface.

SZPs are about equal whether using a blue pigment with a far displacement zeta potential of -40 to -50 mV in 1 mM KNO_3 or an 820 nm polystyrene latex with a far displacement zeta potential of -70 to -80 mV in the same salt solution.

Clearly, it would be useful to find a sample suitable for use with a classic streaming potential instrument and one that would work with an SZP determination. It would be interesting to find out if the same result occurred.

VI-9 Suggested Experiments

What is the effect of variable spreading time? Too short a time should mean some part of the surface is not coated and the average SZP should change with exposure time. But given the relatively small surface area (4.5 mm by 9 mm), this effect may be hard to capture.

What is the effect of different surfactant concentrations? The SZP should change until the entire sample's adsorption sites on the solid surface are occupied. So, plotting SZP vs concentration should yield a negatively sloped line followed by a flat line. The concentration at which the line changes slope is the saturation concentration. Adding more surfactant does not increase the surface zeta potential. It only wastes surfactant.

What is the effect of using different surfactants? There should be different SZP values at the same concentration indicating charge density added to the surface. The economics of one surfactant over another can be examined. And, of course, the health and environmental effects of one surfactant vs. another can be explored.

Aging, rinsing, and additional surface treatments can all be explored regarding maintaining, decreasing, or increasing the SZP value.

Chapter VII: Sedimentation & Centrifugation: d_p, PSD

VII-1 Resolution

If a measurement of size can distinguish a 100-nm particle from a 200-nm particle, the resolution is said to be at least 2:1. In Chapter III on DLS, recall that this is approximately the limit on resolution for that technique, though it depends on the ratio of the scattered intensity of each size and how smooth the autocorrelation function is. It is sometimes possible to distinguish sizes more closely, but not routinely. For this reason, DLS may be called a medium resolution technique. It is one example of an ensemble averaging instrument. The raw data in such instruments contains both size and relative amounts (either by intensity, volume, surface area or number) in one function such as the autocorrelation function, or, in the case of laser diffraction, the angular intensity pattern. Deconvolution is then required to produce a size distribution. Such deconvolution limits resolution.

The particle size distribution, PSD, consists of both size and amount information.

VII-2 Low Resolution: Just One Moment Please

What is a low-resolution particle sizing technique? One that produces just an average size, the first moment of the distribution, is an example. Recall that from the measurement of surface area per unit mass discussed in Chapter I, the surface area average diameter could be determined $<d>_{32}$, but nothing about the width of the distribution or how many modes it might have.

Another example,[44] also involving the specific surface area average diameter,$<d>_{32}$, is shown on the next page.

---

[44]"Determination of Specific Surface Area of Colloidal Silica by Titration with Sodium Hydroxide", G.W. Sears, Anal. Chem. vol. 28 no. 12 pp 1981-1983, (1956). Sears used the fact that at pH 9, 1.26 hydroxyl ions are absorbed per nm^2 of surface on colloidal silica.

Low Resolution Particle Sizing: Specific Surface Area using Titration

To neutralize the hydroxyl groups on 1.751 g of silica required 4.97 mL of 0.1 M HNO_3 acid. Given 1.26 nm² of surface area average diameter, $<d>_{32}$.

0.1 M/L x 4.97 mL = 4.97 x 10^{-4} moles H^+ added. Therefore 4.97 x 10^{-4} moles of OH^{-1} groups were present. Assuming all were on the surface of the particles, there must be 1.26 nm²/mole OH^{-1} ion · 4.97 x 10^{-4} moles OH^{-1}· 6.022 x 10^{23} OH^{-1} ion/mole = 3.77 x 10^{20} nm² of total surface area, S_{tot}. Therefore, the Specific Surface Area, SSA = 3.77 x 10^{20} nm²/1.751 g = 2.15 x 10^{20} nm²/g.

Since SSA = $6/\varrho·<d>_{32}$, it follows that $<d>_{32}$ = 6/(2.196 g/cm³·2.15 x 10^{20} nm²/g) = 1.27 x 10^{-20} cm³/nm² x 10^{+21} nm³/cm² = 12.7 nm.

Summary: From titration data, the surface area average diameter was calculated.

Batch mode static light scattering, SLS, offers another example. The following are produced: $<M>_w$, the weight-average molecular weight and $<R_g>_z$, the z-average radius of gyration. One obtains a single, average size that is descriptive of the distribution but says nothing about its breadth. (SLS, when used in flow mode is capable of reconstruction of the full distribution as in size exclusion chromatography, SEC, also known as gel permeation chromatography, GPC.)

Sometimes only one moment of the distribution is known. And sometimes that is all that is necessary to characterize one sample from another. It depends on how much information is required. So, don't dismiss low resolution methods entirely.

VII-3 High Resolution: Single Particle Counters, SPCs

The ultimate in resolution is to count every single particle and size it. This is the job of a single particle counter (SPC). If you know the volume of the liquid too, you have absolute concentrations in each size class. Blood counters, micro-contamination counters (clean rooms), and image analyzers are all SPC devices. In principle, they yield the highest resolution and offer the possibility of yielding absolute concentrations. In addition, image analysis offers shape information.

SPCs have drawbacks. A little statistical analysis (See Appendix S1) shows that for broader distributions, one needs to count a lot of particles in each size class. This is rarely done. Then there is the problem of coincidence counting when two or more particles contribute to a single signal. The way out of this problem is to dilute to prevent coincidence counting; yet, this only puts a greater burden on being able to count enough particles to reduce statistical error. Some SPCs count particles as they create signals when passing through an orifice such as an electro-zone counter or its optical analogue. Here a new problem arises clogging.

Still, if a distribution is very narrow or consists of narrow peaks, and the concentration is sufficiently low to prevent coincidence counting, and the orifice (if one is employed) is chosen properly, a SPC offers the highest resolution. Consider this: For a truly monodisperse sample, only one particle must be measured; for two peaks, each monodisperse, a relatively few particles of each size need be measured to establish size and amount in each peak. Thus, SPC devices work best with narrow distributions of one or a few peaks. They are least useful with broader distributions.

VII-4 Fractionation Techniques—Good to High Resolution

In between the single particle counters and the ensemble averaging instruments lies the fractionation instruments. These includes sedimentation and centrifugation instruments, sieves, as well as field-flow and column hydrodynamic fractionation devices. With these techniques, the distribution is physically separated into bands and the size and amount are determined by different measurements. In the remainder of this chapter, the first of these types—sedimentation and centrifugation instruments—will be explored in detail.

VII-5 Sedimentation—A Rock in A Pond

Imagine a particle of mass m_p and density ρ_p at the surface of a pond with density ρ_L and viscosity η_L. What are the forces on the particle? There are three:

Gravity pulls the particle down— $+m_p \cdot g$
Buoyancy resists the downward pull— $-m_L \cdot g$
Friction, proportional to velocity, also resists the downward pull— $-f_D \cdot (dx/dt)$

Per Newton's 2nd Law of Motion, the sum of the three forces must be equal to particle mass times its acceleration, d^2x/dt^2, where x is positive pointing down:

$$m_p \frac{d^2x}{dt^2} = m_p g - m_L g - f_D \frac{dx}{dt} \qquad \text{(VII-1)}$$

This is a linear, 2nd order, homogeneous differential equations with constant coefficients. As such, given limiting conditions (x = 0 and $\dot{x} = 0$ when t = 0), it has an exact solution. (See Appendix S2). But under practical conditions, measurements of at least a few minutes, the terminal velocity (constant velocity) is reached in a few milliseconds, leads to Stokes Law of Sedimentation under gravity. When the velocity, dx/dt, is constant, the acceleration, $d^2x/dt^2 = 0$, and equation VII-1 reduces to:

$$\frac{dx}{dt} = \frac{(m_p - m_L)g}{f_D} \qquad \text{(VII-2)}$$

Integration yields Stokes' Law of Sedimentation as follows:

$$\int_0^h dx = \frac{(m_p - m_L)g}{f_D} \int_0^t dt \qquad \text{(VII-3)}$$

where h is the distance traveled. The difference in mass is equal to the difference in density, $\Delta\rho = \rho_p - \rho_L$, times the volume of the particle, v_p, leading to:

$$t = \frac{h \cdot f_D}{g \cdot \Delta\rho \cdot v_p} \qquad \text{(VII-4)}$$

Up to this point no particle shape has been assumed. For a sphere, $v_p = (\pi/6)d_p^3$, and the frictional coefficient for a sphere is given by Stokes' law of friction as:

$$f_D = 3\pi \eta_L d_p \qquad \text{(VII-5)}$$

This is the same equation used in DLS. Thus, the diameter determined is called the hydrodynamic diameter, which includes the particle and whatever moves with it such as surfactants, dispersing agents, ions and solvent, most of which are a small fraction of the particle size. In the case of Stokes' Law of Sedimentation, it is also called Stokes' diameter, d_{St}. It is an equivalent spherical diameter, the diameter of a sphere that settled at the same rate as the actual particle, which may not have been a sphere.

Using equations VII-4 and VII-5, and the volume of a sphere, one obtains the time it takes to travel a distance h, as given by Stokes Law of Sedimentation:

$$t = \frac{18 \cdot \eta_L \cdot h}{g \cdot \Delta\rho \cdot d_p^2} \qquad \text{(VII-6)}$$

The following cartoon, Figure VII-1, illustrates the situation.

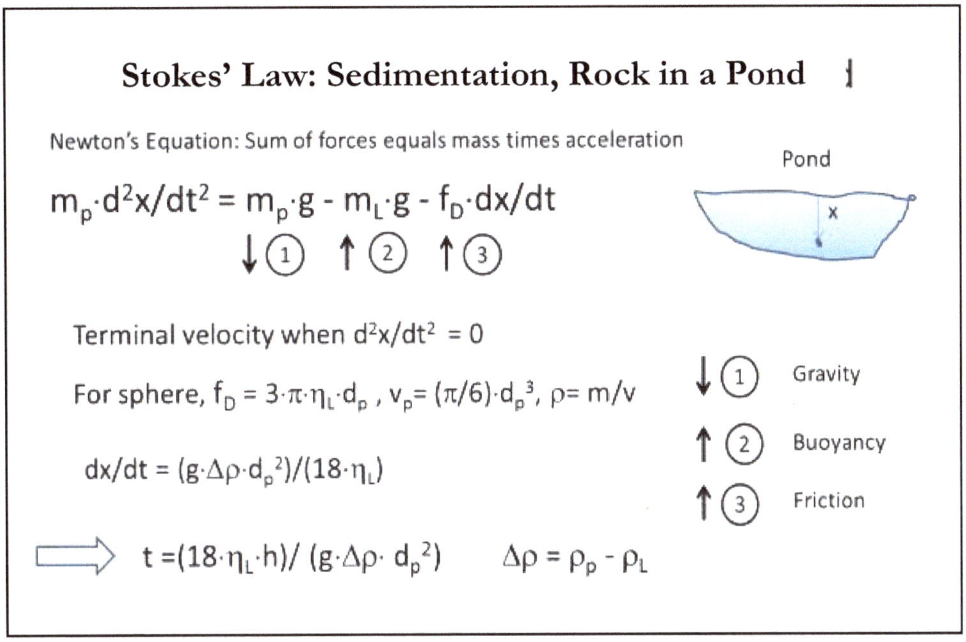

Stokes' Law: Sedimentation, Rock in a Pond

Newton's Equation: Sum of forces equals mass times acceleration

Pond

$$m_p \cdot d^2x/dt^2 = m_p \cdot g - m_L \cdot g - f_D \cdot dx/dt$$

$$\downarrow ① \quad \uparrow ② \quad \uparrow ③$$

Terminal velocity when $d^2x/dt^2 = 0$

For sphere, $f_D = 3 \cdot \pi \cdot \eta_L \cdot d_p$, $v_p = (\pi/6) \cdot d_p^3$, $\rho = m/v$

$dx/dt = (g \cdot \Delta\rho \cdot d_p^2)/(18 \cdot \eta_L)$

$\downarrow ①$ Gravity

$\uparrow ②$ Buoyancy

$\uparrow ③$ Friction

$$\Longrightarrow \quad t = (18 \cdot \eta_L \cdot h)/(g \cdot \Delta\rho \cdot d_p^2) \qquad \Delta\rho = \rho_p - \rho_L$$

Figure VII-1: Stokes' Law of Sedimentation under Gravity.

VII-6 Consequences of Stokes' Law of Sedimentation

For liquids of high viscosity, for smaller differences in liquid and particle density, placing the detector at a larger depth (larger h), and for small particles, the sedimentation time is longer. For liquids of low viscosity, for larger differences in liquid and particle density, placing the detector at a smaller depth (smaller h), and for large particles, the sedimentation time is shorter. Look at a few examples.

The acceleration due to gravity $g = 981$ cm/s^2. Assume the detector is placed at a depth of h = 1 cm and the liquid is water at 25 °C, so $\rho_L = 0.997$ g/cm^3 and $\eta_L = 0.890$ mPa·s = 0.890 x 10^{-2} g/cm·s. Pick three cases with a low, medium and high density difference with water: poly(methyl methacrylate) latex, PMMA, $\rho_p = 1.185$ g/cm^3; silicon dioxide, SiO$_2$, $\rho_p = 2.65$ g/cm^3; and lead, Pb, $\rho_p = 11.34$ g/cm^3. Results shown in Table VII-1.

Time to sediment 1 cm			
Size (μ)	PMMA	SiO$_2$	Pb
1	24 hr	2.7 hr	26 min
5	58 min	6.6 min	63 s
10	14 min	100 s	16 s
50	35 s	4 s	0.6 s

Table VII-1: Sedimentation times under gravity using Stokes' Law.

As expected, denser, larger particles sediment faster. The numbers show more. When the time is too large (PMMA, 1 μ & 5 μ; SiO$_2$ 1 μ; Pb 1 μ) compared to other methods, gravitational sedimentation is not convenient for particle sizing. One trick is to move the detector, so h is smaller, but accuracy will eventually suffer. When the time is too small (SiO$_2$ & Pb 50 μ; possibly Pb 10 μ), accuracy may suffer in timing the detector crossing. A trick here is to increase the liquid viscosity. This is often done for large, dense particles using mineral oil. Cleaning up afterwards is an inconvenience.

Perhaps the biggest problem is with broad size distributions. From Stokes Law, if the ratio of the largest to smallest size is 10:1, the ratio in detector crossing time is 1:100. Arranging the largest to cross in 1 min means the smallest crosses in 100 min. This is a direct result of $t \propto 1/d^2$. One trick here is to scan the detector at constant speed and calculate h from the liquid meniscus to the detector position as a function of time. This can collapse the time ratio by perhaps a factor of three or four but any faster and resolution, the prime benefit of sedimentation, increasingly suffers. Placing detectors at various positions (furthest away registers largest particles in a reasonable time; closest to the meniscus registers smallest particles in a reasonable time; perhaps a third detector to register the middle of the distribution in a reasonable time) is another possibility. But different detectors have different responses and must be calibrated properly and from time-to-time.

When particles are less dense than the liquid, $\Delta\rho < 0$, they rise not fall. This is called creaming. The detector is place just below the meniscus and the particles are initially homogeneously dispersed throughout the liquid. Particles of varied sizes will cross the detector at the same time depending on their starting positions. The math describing this technique, called the HOST technique (homogeneous start technique is referenced later. The HOST technique may also be used when $\Delta\rho > 0$.

For sizes below a few microns, and especially for low density particles, the best trick is to increase the force by using centrifugation. As will be shown below, g is replaced by $\omega^2 r$, where ω is the rotational speed of the spinning disc in which the particles sediment radially outward (or inward if $\Delta\rho < 0$, for particles less dense than the liquid). The math in equation VII-6 describes the LIST technique (line start technique) where all the particles start from the same position at the meniscus.

VII-7 Centrifugation—A Rock in A Spinning Pond

For centrifugation, imagine a spinning, hollow disc with a cylindrical cavity into which a volume of liquid is injected. This is called the spin fluid. Ignoring for now the gradient that is usually present and a very thin layer of liquid injected to prevent evaporative cooling, at the start of a LIST measurement (all particles starting at the "line" constituted by the surface of the spinning liquid), a small volume of a dilute suspension of particles is injected and centrifugal separation begins. If the particle has mass m_p and density ρ_p, and the spin fluid density ρ_L and viscosity η_L, what are the forces on the particle? There are three:

Centrifugal pulling particle radially outward— $+m_p \cdot \omega^2 \cdot r$

Buoyancy resisting the outward pull— $-m_L \cdot \omega^2 \cdot r$

Friction, proportional to velocity, also resists the outward pull— $-f_D \cdot (dr/dt)$

Here ω is the rotational speed of the disc in radian/second and r is the radial distance from the center of the disc.

Per Newton's 2nd Law of Motion, the sum of the three forces must be equal to particle mass times its acceleration, d^2r/dt^2, where r is positive pointing radially outward:

$$m_p \frac{d^2 r}{dt^2} = m_p \omega^2 r - m_L \omega^2 r - f_D \frac{dr}{dt} \qquad \text{(VII-7)}$$

This is a linear, 2nd order, homogeneous differential equations with constant coefficients. As such, given limiting conditions ($r = R_s$, radius of the meniscus, and $\dot{r} = 0$ when $t = 0$), it has an exact solution. (See Appendix S2). But under practical conditions, measurements of at least a few minutes, the terminal radial velocity (constant velocity) is reached in a few milliseconds, leads to Stokes' Law of Centrifugation. When the velocity, dr/dt, is constant, the acceleration, $d^2r/dt^2 = 0$, and equation VII-7 reduces to:

$$\frac{dr}{dt} = \frac{(m_p - m_L)\omega^2 r}{f_D} \qquad \text{(VII-8)}$$

Integration yields Stokes Law of Sedimentation as follows:

$$\int_{R_s}^{R_D} dr = \frac{(m_p - m_L)\omega^2 r}{f_D} \int_0^t dt \qquad \text{(VII-9)}$$

Here R_D is the radial position of the detector and R_s is the initial radial position of the particles, which is the spinning meniscus. See Figure VII-2. These values are determined by the dimensions of the disc, the injected spin fluid volume, and the position of the detector as will be shown later.

The difference in mass is equal to the difference in density, $\Delta\rho = \rho_p - \rho_L$, times the volume of the particle, v_p, leading to:

$$t = \frac{Ln(\frac{R_D}{R_s}) f_D}{\omega^2 \Delta\rho v_p} \qquad \text{(VII-10)}$$

Up to this point no particle shape has been assumed. For a sphere, $v_p = (\pi/6) \cdot d_p^3$, and the frictional coefficient for a sphere is given by Stokes' law of friction as:

$$f_D = 3\pi \eta_L d_p \qquad \text{(VII-11)}$$

This is the same equation used in DLS. Thus, the diameter determined is called the hydrodynamic diameter, which includes the particle and whatever moves with it such as surfactants, dispersing agents, ions and solvent, most of which are a small fraction of the particle size. In the case of Stokes Law of Centrifugation, it is also called Stokes' diameter, d_{St}. It is an equivalent spherical diameter, the diameter of a sphere that settled at the same rate as the actual particle, which may not have been a sphere.

Using equations VII-10 and VII-11, and the volume of a sphere, one obtains the time it takes to travel a distance R_D - R_S, as given by Stokes Law of Centrifugation:

$$t = \frac{18\eta_L Ln(\frac{R_D}{R_S})}{\omega^2 \Delta\rho \cdot d_p^2} \qquad \text{(VII-12)}$$

The following cartoon, Figure VII-2, illustrates the situation.

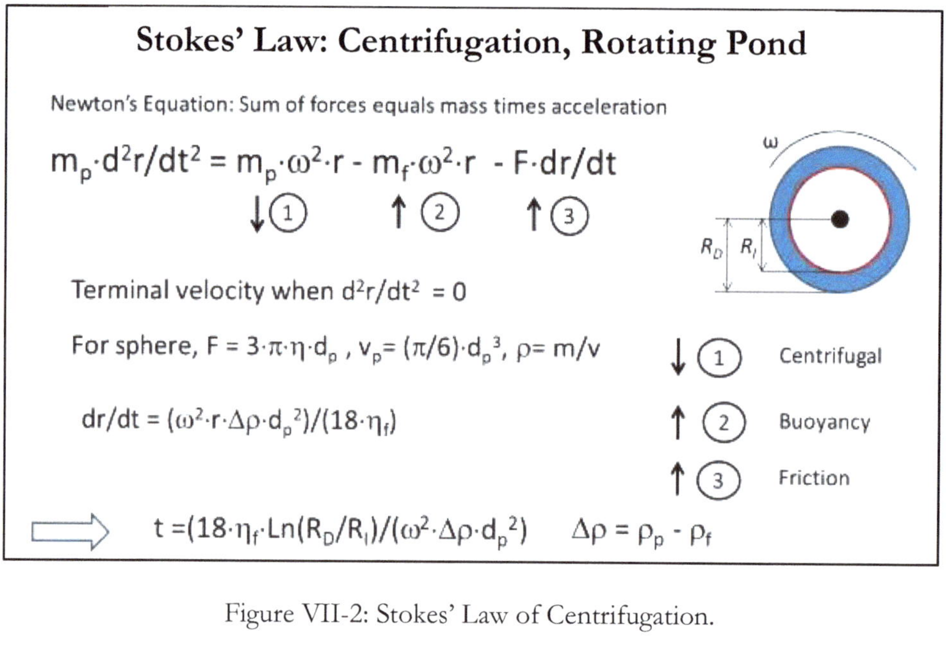

Figure VII-2: Stokes' Law of Centrifugation.

The results for sedimentation equation VII-6 and centrifugation equation VII-12, indeed, the equations to get there, are perfectly analogous. Both results depend on η_L, $\Delta\rho$, and d_p^2 in the same way: Particles reached the detector faster in low viscosity liquid, with higher density differences, and with larger particle sizes. And they reached the detector more slowly in high viscosity liquid, with low density differences, and with smaller sizes. The difference is that g is replaced by ω^2 and h by $Ln(R_D/R_s)$. These changes will allow much smaller sizes to be reached in a reasonable time but put upper limits on the largest and smallest sizes. See Appendix S3 and S4.

VII-8 Consequences of Stokes' Law of Centrifugation

For liquids of high viscosity, for smaller differences in liquid and particle density, using a larger spin fluid volume (making the difference between R_D and R_s larger) and for small particles, the centrifugation time is longer. For liquids of low viscosity, for larger differences in liquid and particle density, using a smaller spin fluid volume (making the difference between R_D and R_s smaller), and for large particles, the centrifugation time is shorter. Look at a few examples.

Pick three cases with a low-, medium- and high-density difference with water or, in the case of AgCl, a 24% w/w ethylene glycol-water solution: poly(methyl methacrylate) latex, PMMA, $\rho_p = 1.185$ g/cm³; silicon dioxide, SiO_2, $\rho_p = 2.65$ g/cm³; and silver chloride, AgCl, $\rho_p = 5.56$ g/cm³. The relationship between R_S (R_l in Figure VII-2), the disc dimensions, and spin fluid volume (12, 15, or 20 mL) is given in equation VII-14. Centrifugation times shown in Table VII-2.

Time to reach detector in minutes under specified conditions						
Size (nm)	PMMA	Disc Conditions	SiO_2	Disc Conditions	AgCl	Disc Conditions
1,000	3.95	15 mL, 2,000 rpm	2.89	20 mL, 1,000 rpm	2.46	20 mL, 1,000 rpm
500	3.95	15 mL, 4,000 rpm	2.89	20 mL, 2,000 rpm	2.46	20 mL, 2,000 rpm
100	15.79	15 mL, 10,000 rpm	2.89	20 mL, 10,000 rpm	9.57	15 mL, 4,000 rpm
50	29.48	12 mL, 12,000 rpm	7.18	15 mL, 10,000 rpm	9.57	15 mL, 8,000 rpm

Table VII-2: Sedimentation times under centrifugation using Stokes' Law.

As expected, denser, larger particles centrifuge faster. The numbers show more.

Notice that by doubling the rotational speed when the size is halved keeps the crossing time the same. Notice also that the times vary from a few minutes to almost 30 min. This is typical and is generally much longer than a single DLS measurement. When applicable, this longer time is compensated by an increase in resolution, a direct result of particle separation prior to detection.

The calculations were done for a single size. A bigger problem is with broad size distributions. From Stokes' Law of Centrifugation, if the ratio of the largest to smallest size is 10:1, the ratio in detector crossing is 1:100: arranging the largest to cross in 20 s means the smallest crosses in 2,000 s (33 min), emphasizing again the longer time compared to, for example, DLS. This is a direct result of $t \propto 1/d^2$. One trick here is to scan the detector at constant speed and calculate R_D from the liquid meniscus to the detector position as a function of time. This can collapse the time ratio by perhaps a factor of three or four but any faster and resolution, the prime feature of centrifugation, increasingly suffers.

VII-9 Relationship Between Disc Dimensions, Spin Fluid Volume, and Meniscus Radius R_S

Assume the disc contains a cylindrical cavity of width w and cavity radius R_c. Its volume is therefore $\pi R_c^2 w$. When a spin fluid volume V_{SF} is injected, and the disc spun, a cylindrical cavity of air is created with radius R_s and width w. The following relationship holds:

$$\pi R_c^2 w = \pi R_S^2 w + V_{SF} \tag{VII-13}$$

From this it is easy to calculate R_S as follows:

$$R_S = \sqrt{R_c{}^2 - \frac{V_{SF}}{\pi w}} \tag{VII-14}$$

For example, a typical disc might have dimensions of w = 0.6340 cm and R_C = 5.0720 cm determined using a cathetometer and accurate volumetric pipettes of two different volumes. Measuring the diameter ($2R_S$) of the meniscuses when separately 5.00 mL and then 20.00 mL are injected and spinning, for example, allows one to calculate R_C and w using equation VII-13. A convenient place for the detector is at the meniscus with just 5.00 mL spinning fluid volume. It is just above the inner disc circumference. Thus, R_D is known from the cathetometer data. A value associated with the same disc described above is R_D = 4.8182 cm.

If V_{SF} = 20.0 mL, the value of R_S from equation VII-14 is calculated as follows:

$$R_S = \sqrt{5.0720^2 - \frac{20.0}{\pi \cdot 0.6340}} = 3.9603 \ cm \tag{VII-15}$$

We are now able to calculate one of the entries in Table VII-1. For 100 nm SiO_2 using V_{SF} = 20.0 mL of water at 25 °C, with the disc just described rotating at 10,000 rpm (1,047.2 rad/s), equation VII-12 yields for the crossing time:

$$t = \frac{18 \cdot (0.890 \cdot 10^{-2}) \cdot Ln(\frac{4.8182}{3.9603})}{1,047.2^2 \cdot (2.65 - 0.997) \cdot (100 \cdot 10^{-7})^2} = 173.3 \ s = 2.89 \ min \tag{VII-16}$$

VII-10 Measurement Conditions

It was discovered long ago that injecting a suspension into a spinning disc with fluid too often led to failure where the particles did not centrifuge radially outward (for $\Delta\rho > 0$) in accordance with Stokes' Law. A gradient at the meniscus was required for smooth entry of the particles. The gradient can be achieved in a variety of ways, but here is a useful method. Inject first a few tenths of a milliliter of a liquid with lower viscosity and density than the spin fluid. Start the disc rotating and immediately inject the spin fluid. Due to centrifugal forces, the gradient liquid rises through the spin fluid before it can mix homogeneously. This forms a very thin layer at the meniscus and can be ignored in the calculation. To avoid evaporative cooling, about 0.1 mL of, typically, dodecane, is injected. Its lower density than either the gradient liquid or spin fluid ensures it remains on top. Its lower vapor pressure ensures little or no evaporation and the cooling of layers below that would have followed. If significant cooling did occur, the layers below drop in temperature, increase in viscosity, and particles are retarded. In the worst case, a denser and more viscous layer of spin fluid riding on top of warmer and less dense and viscous layers below can result in hydrodynamic instability, a form of uncontrolled mixing. This results in complete disruption of the expected particle flow. They no longer follow Stokes' Law. The detector signal typically levels off.

VII-11 Detection: Disc Centrifuge Photosedimentometer, DCP

Particles scatter and for some, absorb, light. Therefore, a useful detector is a beam of light shot through the disc close to the inner circumference. A narrow slit is used to define the beam, but the slit width has consequences. But first, consider why light detection is useful as shown in Figure VII-3.

Why Light Detection?

- $I_t = I_o \cdot \exp(-\tau \cdot w)$ I_o, I_t = Transmitted, Incident Intensity

- $\tau = Q_{ext} \cdot c$ = turbidity

- Q_{ext} = Extinction efficiency = $f(d_p, n_p/n_o, \lambda)$ Mie calculation

- c = Mass concentration

- w = Path length (width of liquid across disc)

- Light is either absorbed or scattered, the sum of which is called extinction. Extinction is a strong function of particle size. *Significant optical corrections are necessary for $d_p \leq 5\,\mu$.*

- Sensitive but less quantitative for high density materials where Q_{ext} not easy to calculate accurately.

Figure VII-3: Light detection used in a DCP for particle sizing.

The signal at any time is proportional to I_t, the transmitted intensity, and the signal at the baseline is proportional to I_o, the incident intensity. The ratio is called the transmittance. Its logarithm is the product of the extinction efficiency, which can be calculated from Mie theory for spheres, the mass concentration c of the particles crossing at time t, and the pathlength cross the disc, w.

The turbidity is calculated from the measurement, but it is the differential volume distribution, dV/dd_p, that is required. (For a distribution of particles with the same density, it is easy to show this is proportional to the differential weight or mass distribution.) The full development is given by Devon, et al.[45] including the correction for a finite slit as well as a radial dilution correction if the detector is scanned. Equation 22 in Devon et al. is repeated here:

$$\tau(d_p) = \left.\frac{\left(\frac{dV}{dd_p}\right) \cdot Q_{ext}}{16 \cdot w \cdot R_D^2 \cdot Ln\left(\frac{R_D}{R_s}\right)}\right. \tag{VII-17}$$

If the detector is not scanned, R_D is a constant and $dV/dd_p \propto \tau/Q_{ext}$. If the detector is scanned, R_D is a function of time and the term $(R_D)^2 \cdot Ln(R_D/R_S)$ must be included to account for radial dilution (increasing concentration if detector scanned up; dilution if detector scanned down for cases where $\Delta\rho < 0$).

Before discussing the consequences of Q_{ext}, it is worthwhile looking at line start (LIST) results.

VII-12 High Resolution Results using DCP

The differential weight (volume) and its cumulative (the integral curve showing steps) distribution for a mixture of four, standard latexes are shown in Figure VII-4.

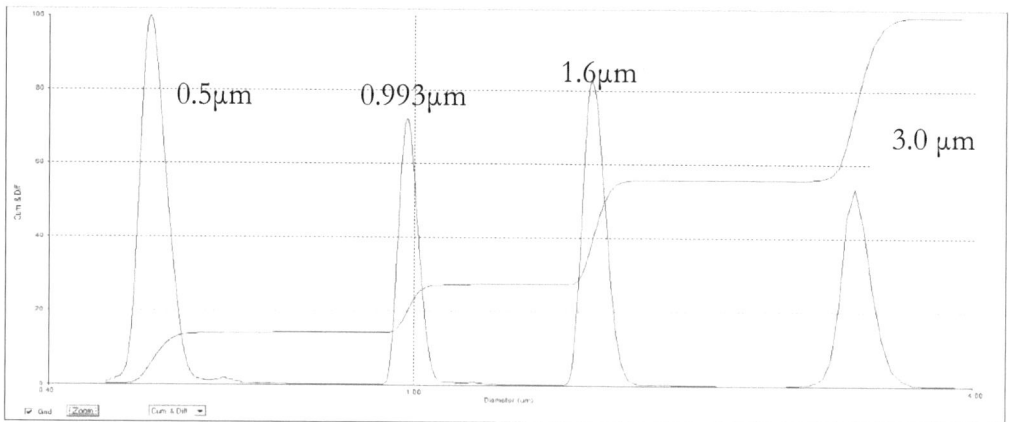

Figure VII-4: Differential and cumulative distributions for mixture of four narrow latexes.

The results agree with the standard values to within 2%. While most practical distributions

---

[45] Devon, M. J., Meyer, E., Provder, T., Rudin, A. and Weiner, B. B., Chapter 10, "Detector Slit Width Error in Measurement of Latex Particle Size Distributions with a Disc Centrifuge", in "Particle Size Distributions II, Assessment and Characterization", ed. Theodore Provder, ACS Symposium 472, (1991).

are not quadrimodal, this is a good demonstration of what high resolution means. Note too that these measurements are absolute in the sense that neither Stokes' Law nor the equations governing the detection have an unknown constant that would entail calibration.

A measure of resolution is the ratio between two narrow peaks as demonstrated in Figure VII-5. With nearly baseline separation, the ratio of nearest neighbor peaks is 1:1.17, only 17% apart. Contrast that with DLS where the analogous figure is 1:2, or 100% apart. Clearly, DCP measurements are much higher resolution.

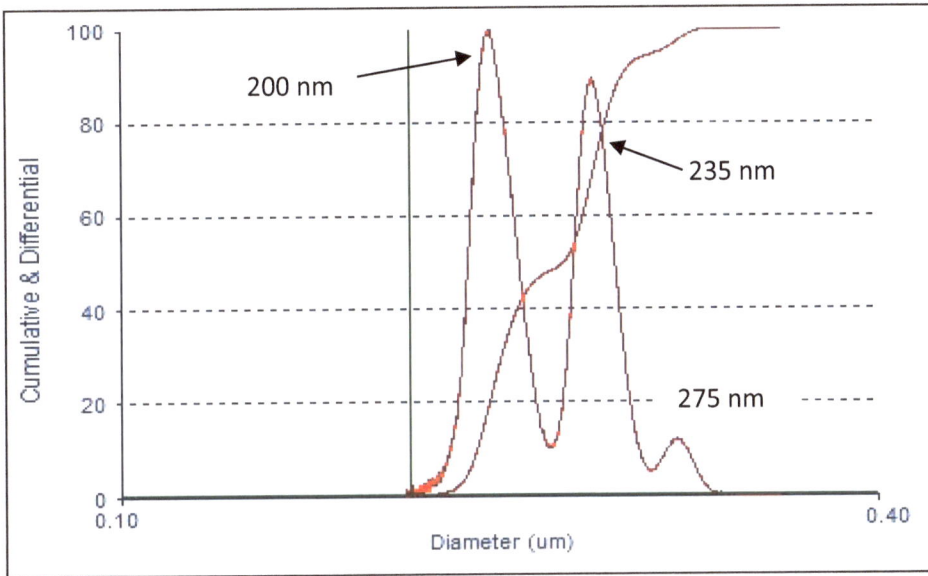

Figure VII-5: DCP measurements demonstrating high resolution.

VII-13 Extinction Efficiency, Q_{ext}

From the raw data of a DCP measurement, if no light scattering correction is applied, a turbidity-weighted distribution is the result. This is not as useful as a volume weighted distribution and so the correction should be applied. It is a function of the wavelength of light, the diameter at which it is calculated, and the refractive indices of particle and spin fluid. The same Mie theory used in calculating angular scattering factors for SLS and DLS is used to calculate Q_{sca}, the extinction due to scattering, and Q_{abs}, the extinction due to absorption.

Absorption is revealed in a non-zero imaginary refractive index. For example, the real part of the refractive index for common glass is about 1.5 and no mention is made of the imaginary part. This is true for visible wavelengths. But in the UV, glass does absorb some of the light (which is why you don't get sunburned in a car), because at those wavelengths the imaginary part is non-zero.

Figure VII-6 shows $Q_{ext} = Q_{sca}$ ($Q_{abs} = 0$) for poly (methyl methacrylate), PMMA, and for polystyrene, PS, both of which do not absorb at the 650 nm wavelength of the narrow-band LED light source used in this DCP to make the measurements. Thus, $n_{im} = 0$ for both materials and n_{re} is displayed in the figure.

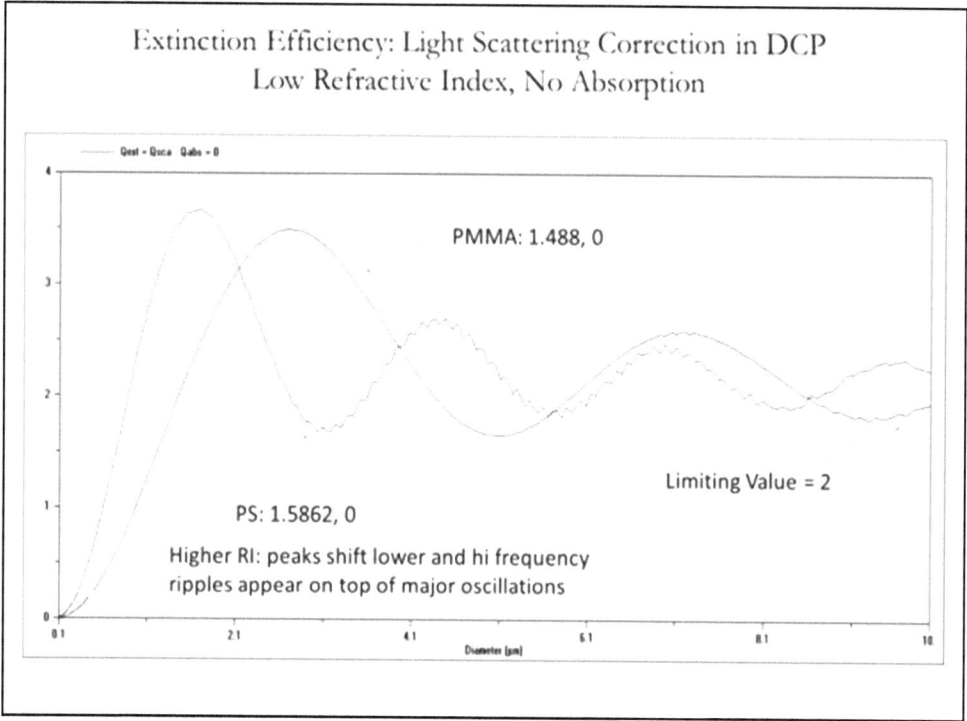

Figure VII-6: Q_{ext} vs d_p for two materials that do not absorb at $\lambda = 650$ nm.

The range on the diameter axis is 0.1 to 10 μ. Both curves show oscillatory behavior but the one with the higher n_{re}, PS, develops high frequency ripples above about 3 μ. High frequency ripples are characteristic, more pronounced, and appear at lower particle size for particles with larger n_{re}. Since the DCP is usually applied to sizing in the colloidal, submicron range, the smoothness of these two curves in that range makes it easy to apply the correction. Notice two other features that are characteristic. First, the higher n_{re}, the lower in size the first peak. Here, for PS it is about 1.5 μ; and for PMMA it is about 2.8 μ. Second, the curves approach a limiting value of 2 for large sizes. This is a classic result from optical theory: Large particles block (extinguish) twice the amount of light that their geometric cross section would predict. That second half is due to scattering around the particle circumference.

Figure VII-7 shows Q_{ext} vs d_p for a metal oxide with a high-real part and a zero-imaginary part of the refractive index. This is one form of titanium dioxide and it is very white, reflecting white light at all visible wavelengths without absorbing in the visible.

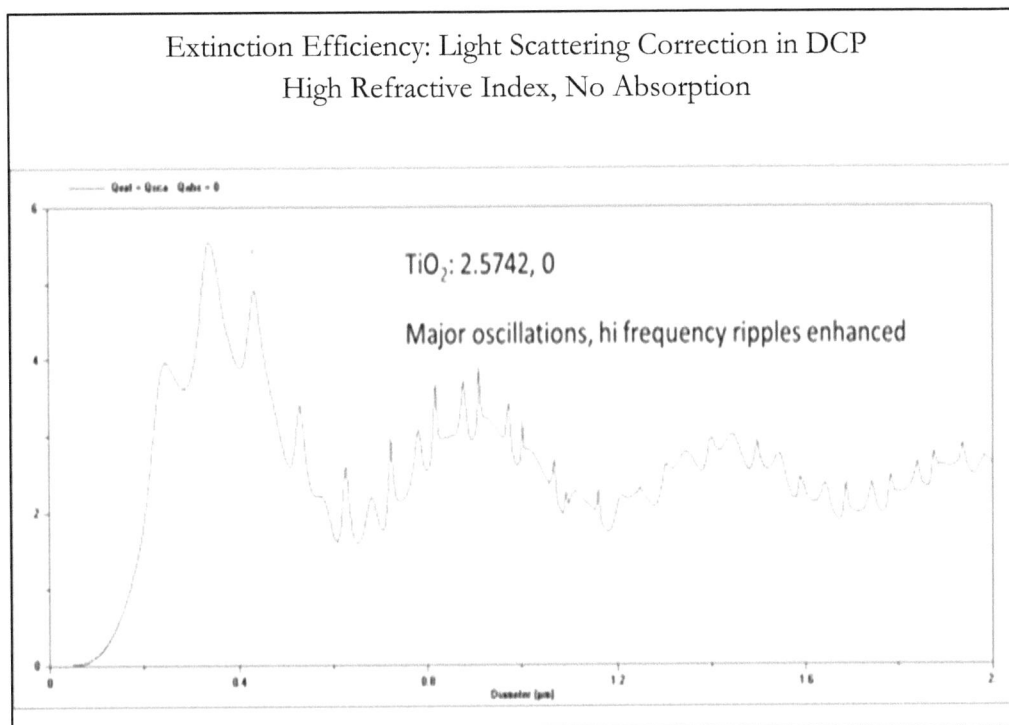

Figure VII-7: Extinction efficiency for TiO_2 vs d_p from 0 to 2 μ.

The characteristic low frequency oscillation is still apparent with the first peak—now split into four sub peaks—upon which the high frequency ripples are superimposed. And the curve is approaching the limiting value of 2 at higher sizes.

Now the size distribution for commercial titanium dioxides used in paints and other applications is a somewhat broad, unimodal distribution ranging from 100 to 500 nm typically, with a peak in the 250-350 nm range. This is right in the region of the sub peaks in Q_{ext}, making it very hard to apply this correction with any accuracy. Thus, the resulting distribution would look choppy, at odds with the real distribution.

Two ways have been developed to deal with these cases. The first is an artificial mathematical trick that does not strictly apply. It leads to smoothing of the ripples. This results in a less choppy size distribution but one that is broadened compared to the real one and, possibly, a shift in the modal value and moments of the distribution.

This same trick is used in some other light scattering techniques, most notably Fraunhofer Diffraction with added high angle detectors. And the same problems may arise.

How was the trick developed? Consider Q_{ext} for a highly absorbing particle, even one with a high real-part to its refractive index: Carbon Black. See Figure VII-8.

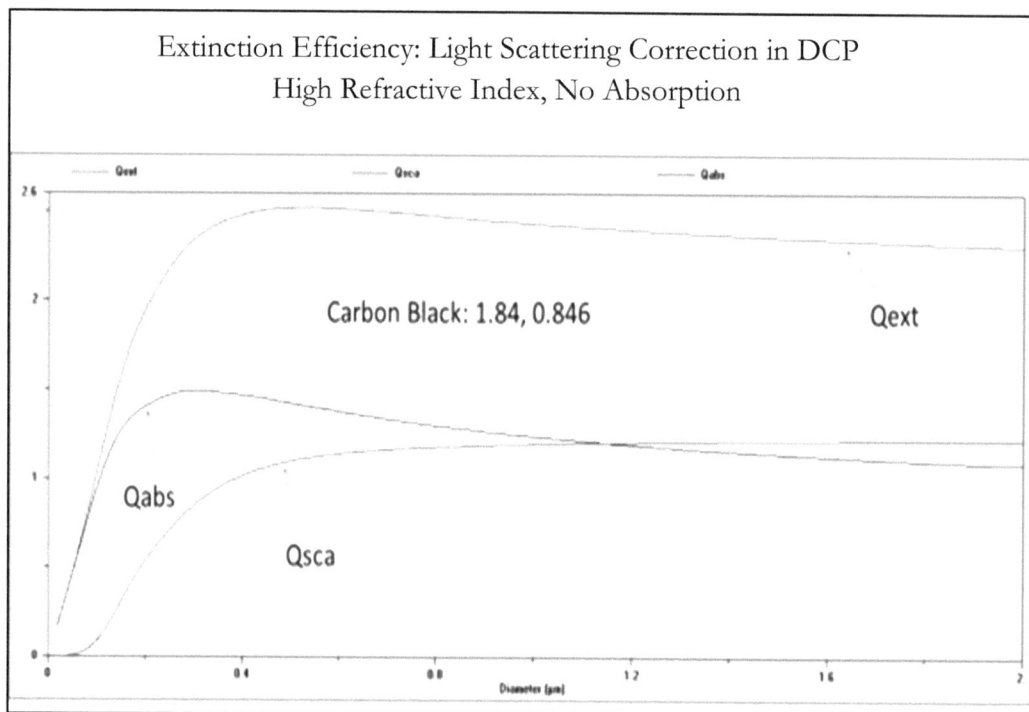

Figure VII-8: Extinction efficiency for Carbon black.

Carbon black is used in tires and as a pigment that produces the color black. It is highly absorbent of visible light which justifies its very high n_{im} = 0.846. Surprisingly, it also refracts any light that passes through as expected by its high n_{re} = 1.84.

Given a high value for the real part, one expects its first peak to be lower than that of either PMMA or PS but higher than that of TiO_2. Indeed, Q_{ext} plateus at about 500 nm consistent with expectation. And it is trending smoothly downward at 2 μ. If the plot were extended to 10 μ, it would be seen that Q_{abs} and Q_{sca} both approach 1, while their sum, Q_{ext} approaches the classic value of 2. What is missing are major oscillations and high frequency ripples.

This suggests a basis for the first trick. By adding a non-zero n_{im}, creating a non-zero Q_{abs}, Q_{ext} is smoother and so are the calculated size distributions.

Figure VII-9 shows an example of this on the next page.

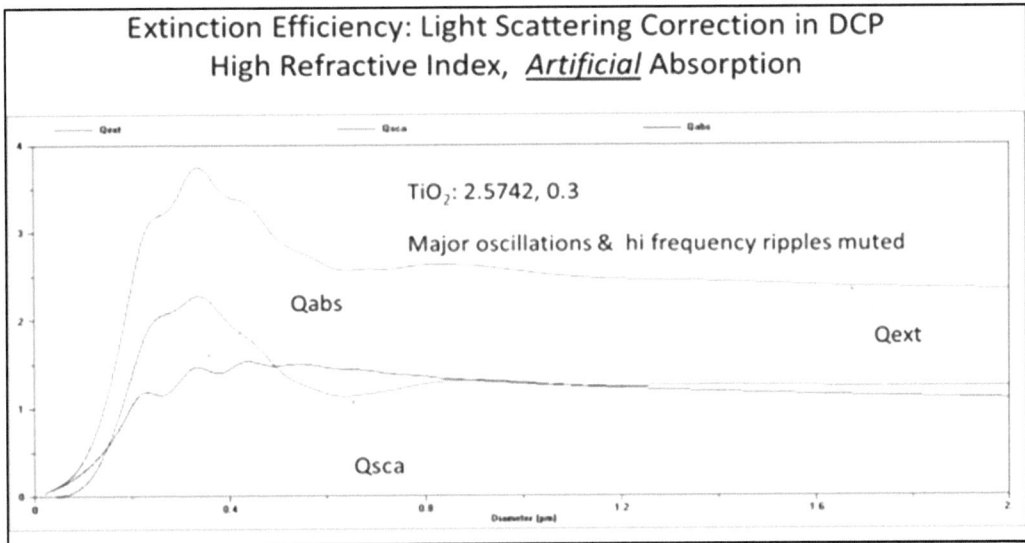

Figure VII-9: Adding artificial adsorption and its effect on Q_{ext}.

Compare the Q_{ext} curve in Figure VII-7 with that in Figure VII-9. The high frequency ripples as well as most of the major oscillations are suppressed. When used as a correction in DCP data, the curve in Figure VII-9 will result in a smoother differential volume distribution as expected. But remnants remain of the modal split of the first major peak. And this was with an arbitrary use of $n_{im} = 0.3$. One could try 0.1, resulting in more ripples and major oscillations; or, one could try 0.5, most likely wiping out the last major oscillation and perhaps muting the remaining modal split considerably. The higher the value the smoother the resulting distribution, but the greater the likelihood of distorting the real distribution.

While the mathematical trick of adding artificial adsorption with a non-zero n_{im} has its uses, there is a better alternative, a better trick: Use a detection technique that doesn't depend on light scattering corrections: Use x-ray detection where applicable.

VII-14 Detection: X-Ray Disc Centrifuge, XDC

High density, inorganic particles are ideal for detection by x-rays. These are the very particles with high refractive indices, with absorption in the visible, where Q_{ext} will exhibit split peaks in the low frequency oscillations and high frequency ripples. Thus, applying the Q_{ext} correction is problematic. Such particles include metal oxides, ceramics, metals and more. Figure VII-10 illustrates why x-ray detection is important for these types of particles.

Why X-Ray Detection?

- $I_t = I_o \cdot \exp(-\mu_{abs} \cdot c \cdot w)$ I_o, I_t = Transmitted, Incident Intensity

- μ_{abs} = X-ray absorption efficiency $\neq f(d_p, n_p/n_f, \lambda)$

- c = Mass concentration

- w = Path length (width of liquid across disc)

- When X-rays are absorbed, they are absorbed in proportion to mass without any dependence on particle size or refractive indices. <u>There is no optical correction.</u>

- For high density materials with normally high refractive indexes, the mass weighted size distribution is quantitative.

Figure VII-10: The value of x-ray detection in particle sizing.

The x-ray absorption efficiency depends on atomic number. For low density, organic particles it is too low to yield a good signal. (Note: Signal limitation is also characterized by the strength of the x-ray source, sources small enough for analytic instrumentation.) Approximately, for atomic numbers greater than 13 (aluminum), the signal is strong enough. Thus, pure aluminum oxide would be difficult to measure.

The result is the concentration $c(d_p)$, which can be transformed into dW/dd_p, the differential weight distribution as a function of d_p. It is assumed all the particles have the same chemical composition so μ_{abs} independent of particle size. The differential weight distribution is sometimes called the differential mass distribution, dM/dd_p, which is proportional to the differential volume distribution, dV/dd_p, if all the particles have the same density independent of particle size.

The HOST (<u>ho</u>mogeneous <u>st</u>art) technique is used for signal strength reasons and because it is difficult to avoid hydrodynamic instabilities with higher density particles. If one could employ a LIST technique, the concentration of particles at the detector would typically not produce a strong enough signal given the power of reasonably sized x-ray sources.

The measurement proceeds initially by establishing an upper baseline where the x-rays are passed through the spin fluid. Then the homogenous suspension, typically about 2% v/v of 15-25 mL, is injected into a stationary disc. With larger sizes and higher particle densities, the disc is rocked back and forth to ensure homogeneity, while the lower baseline is established. Then the disc is made to spin, and it reaches its constant rotational speed in less than a second. The signal is recorded as a function of time.

An example of the raw data is shown in Figure VII-11.

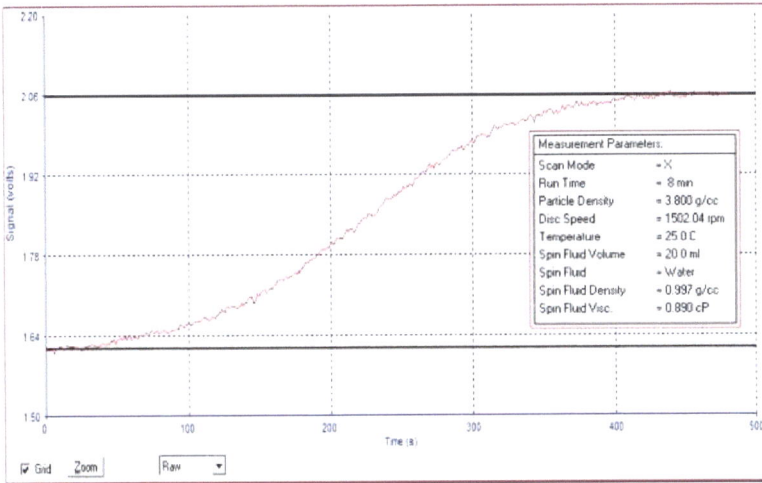

Figure VII-11: Raw XDC data for a form of TiO_2.

Conversion of the raw data to a size distribution is shown in Figure VII-12 and Table VII-3.

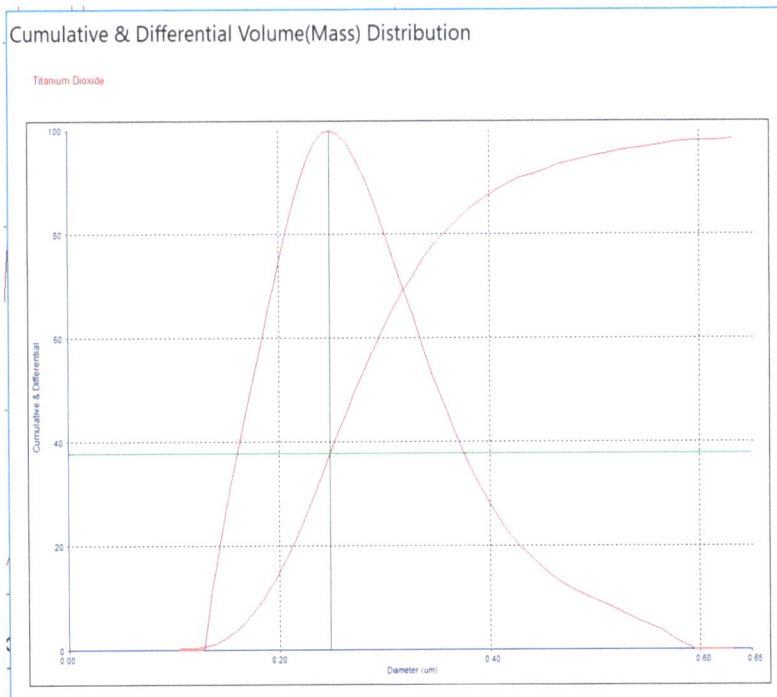

Figure VII-12: Size distribution for TiO_2 using an XDC

Analysis Results : Volume(Mass)			
d10	= 0.186 µm	Mean	= 0.289 µm
d50	= 0.273 µm	Std.Deviation	= 0.095 µm
d90	= 0.420 µm	Mode	= 0.249 µm
Span=(d90-d10)/d50	= 0.858	FWHM	= 0.178 µm
		FWHM/Mode	= 0.714

Table VII-3: Tabular results for TiO_2 using an XDC.

VII-15 Homogeneous Start Technique for Disc Centrifuges, XDC & DCP

When particles are less dense than the liquid, $\Delta\rho < 0$, they rise not fall. This is called creaming. The detector is place just below the meniscus and the particles are initially homogeneously dispersed throughout the liquid. Particles of various sizes will cross the detector at the same time depending on their starting positions. The math describing this HOST technique (homogeneous start) is considerably more complicated than that describing the LIST technique. For that reason, it is not developed here. It is described in some detail in Allen's book[46] and Kamack's seminal paper[47].

The HOST technique may also be used when $\Delta\rho > 0$ and may be used with a DCP. It is always used with an XDC for reasons discussed above.

VII-16 Summary

For high resolution particle sizing, consider sedimentation (higher density, sizes generally above 1 micron) and centrifugation (high and low density, sizes generally below 1 micron). When using centrifugation, use a DCP for low density, typically organic particles. Use an XDC for high density, inorganic particles. When using a DCP, the line start technique yields the highest resolution. In the case where $\Delta\rho < 0$, use the homogeneous start technique with a DCP. The homogeneous start technique is always used with an XDC.

---

[46] T. Allen, "Particle Size Measurements", 3rd edition, Chap. 12, page 371, Chapman & Hall, New York, 1981
[47] H.J. Kamack, *British Journal of Applied Physics,* **5**, 1962-1968, (1972).

APPENDICES

Appendix P1: What Is A Particle?[©]

By Bruce B. Weiner Ph.D., August 2010

Elementary vs Fine Particles: The question embodied in the title might seem strange until you consider the many ways particles play a part in science and engineering. For example, if you do a web search using just the word particle, you will get more hits for atomic, nuclear, and elementary or subatomic particles than you will for pharmaceutical particles, nanoparticles, clay particles, etc. You will get a lot of hits for particle accelerators, particle colliders, particle physics as well as a few hits that really interest you.

It was once proposed by the Fine Particle Society that the term "fine particle" be used as a keyword grouping to avoid confusion with atomic and nuclear particles. The New York State Department of Health defines a fine particle as any particle causing pollution with a diameter of less than 2.5 micron. Such a partial definition undoubtedly arises because of concerns over respiration, but does not cover the lower end, and, of course, it assumes all fine particles cause pollution. They don't. If you do use the search term "fine particle", you usually don't get hits for particle sizers, possibly because the Fine Particle Society in the U.S.A. is not as strong as it used to be, or perhaps because, ultimately, there is no universally agreed upon keywords in this field.

So, the first thing to know about "What Is a Particle?" is to use keywords like these: "particle characterization", which also includes zeta potential and shape; "particle size"; and "particle size analyzer". When you see the term "particle analyzer", do not think it means a chemical analyzer. While misnomer, a particle analyzer is most often an instrument for measuring particle size not a particle's chemical makeup.

The Ideal Particle: From now on when I write "particle", I mean a speck of matter much, much larger than nuclear or subatomic matter, and generally (there are exceptions to everything) larger than small molecules. I am referring to nanoparticles, globular proteins, colloids and larger specks of matter, up into the hundreds of microns.

Now consider what such a particle might look like. Quick, did you picture little, solid spheres mostly of the same size, a monodisperse size distribution? That is the picture most people have when you ask, "What is a particle?" Call it the *ideal particle*. Latex particles used as sizing standards are close to ideal particles in this sense. Just below is a picture of one, though you can't tell if it is truly monodispersed or whether some of the particles are welded together. More on that in another application note about sizing.

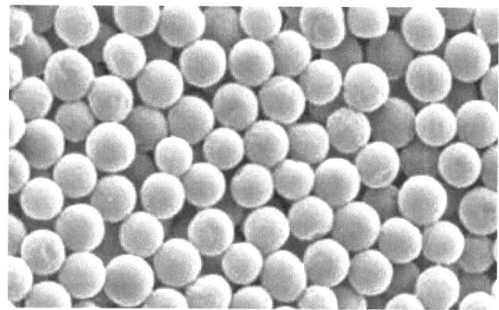

Targeted Drug Delivery with a Spherical Particle: Liposomes, micelles, and surfactant- or polymer-stabilized oil-in-water microemulsions can all be used to hide drugs inside their structures in the hopes that a judicious choice of surface chemistry will guide the carrier to the site of the disease. If a trigger for releasing the highly concentrated and highly specific drug can be activated, medicine is delivered to just the right spot and not throughout the blood stream with its attendant side effects. Alas, the attack of white blood cells and difficulties in finding the right surface chemistry has made the magic bullets only partially successful.

The liposome shown here is cut away to expose the active agents in the interior. The shell of the liposome is most often a bilayer lipid, perhaps 5 nm in thickness. The shell of this liposome is closer to a thin, core-shell than it is to a thick, core-shell particle model. Such models are useful when trying to interpret data for it is often necessary to estimate the refractive index or density of a composite particle such as a liposome.

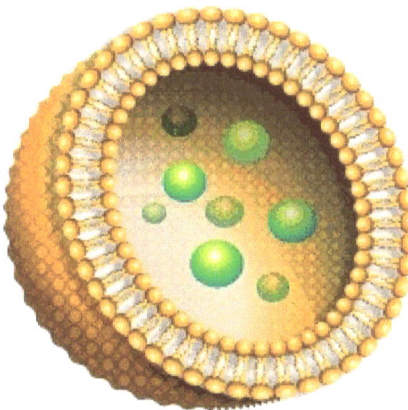

Liposome

The surfactant micelle shown above and to the right is typical. It consists of many surfactant molecules with a polar head (sulphate, for example) and non-polar tails (methyl groups, for example). At a certain concentration in water, the critical micelle concentration (CMC), the surfactant molecules self-assemble into a spherical micelle shape. This reduces the free energy by grouping the non-polar tails together into an oily interior (perfect for trapping not only water insoluble medicines but also greasy or oily dirt, thus forming the basis for detergency).

At higher concentrations, other shapes form. In addition, polymers with a polar head and a nonpolar tail can also form micelles, though far fewer are needed to do so, compared to surfactant micelles.

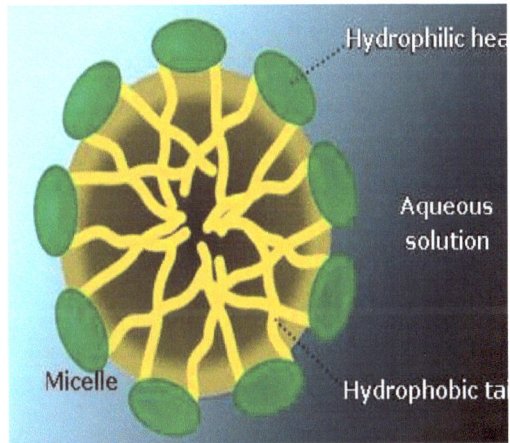

Micelle

Compared to a solid latex particle, neither a liposome nor a micelle is solid. Indeed, when a micelle is filtered, the shear forces can break it apart into individual molecules which then reassemble spontaneously downstream of the filter: self-assembly particle. A filtered liposome, once broken up, does not form again into a spherical particle.

143

Spherical Agglomerates: The calcium carbonate particle shown here is perhaps 35 micron in diameter and it is clearly made up of a lot of smaller, non-spherical building blocks that may or may not be primary aggregates.

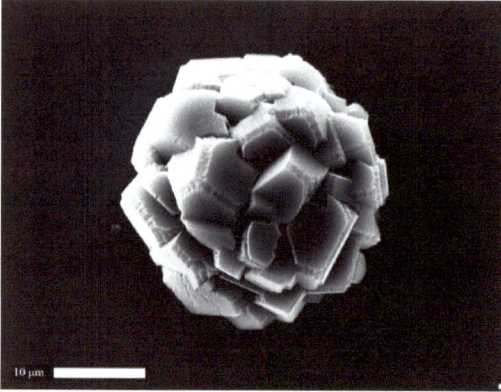

Agglomerate of calcium carbonate.

The blocks are held together by a very large number of attractive forces but not by covalent chemical bonds. When this particle is rotating, it is easy to imagine a nice globe defining its circumference. Such particles are called globular.

Other Globular Particles: The math that defines a sphere's volume or surface area is trivially easy; the math that describes its behavior in a gravity or centrifugal field requires simple Newtonian physics and a few lines of calculus; and the math that defines its interaction with light, while difficult, was solved about 100 years ago and is, thankfully, available on a computer. For these reasons, any particle that resembles a sphere, as it will when rapidly rotating in a liquid at low to modest viscosity, will be called globular. From globular to spherical is one easy conceptual step away and then the size is defined by a diameter or radius.

Globular proteins are single, macromolecules with thousands of covalent and hydrogen bonds that allow the folded structure (tertiary structure) to be much more compact than a linear macromolecule. Note that other particles are made up of lots of individual molecules, but the single protein particle is a single molecule, though sometimes made of tightly bound subunits.

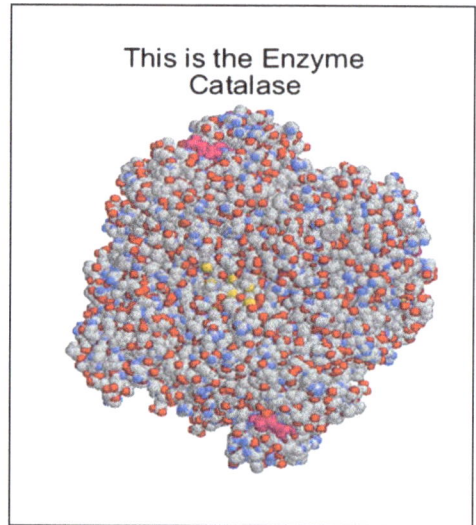

Globular Protein

A 15 nm gold sol particle suspended in a liquid is made up of lots of gold atoms that are not covalently bonded. The latex particle, made using emulsion polymerization, is made up of a lot of long-chain, polymer molecules that are stuck together. The particle is not a single molecule. Likewise, micelles and liposomes are not single molecules.

The image of the red blood cell shown here (unfortunately in black-and-white) has the typical biconcave disc shape. Its disc diameter is, on average, about 7 micron. The larger, 12-15 micron, but also globular white blood cell is shown. Its surface is studded with pseudo-pods.

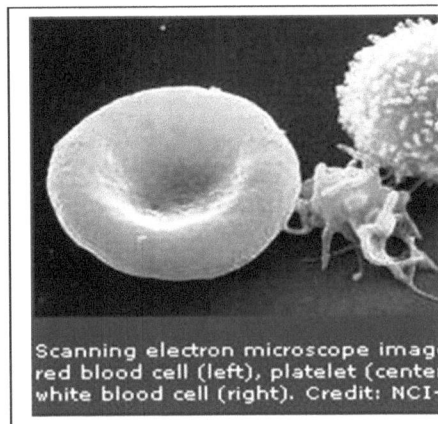

Scanning electron microscope image red blood cell (left), platelet (center white blood cell (right). Credit: NCI-

A highly non-spherical particle, a blood plasma platelet, is also shown. Its resemblance to an alien-invading insect is obvious and it is hard to call it globular. Perhaps its cousins look more disc-like and plate-like to deserve the name platelet. None of these particles are in relative scale. If they were, the RBC would be half the size of the WBC.

Non-spherical Particles, Rods: Not all particles are globular. Some pigments, even some molecules large enough to be called particles, and especially some primary aggregates and loosely-held agglomerates can become elongated enough to be called rods. Though there is no exact length-to-width ratio (called the aspect ratio) that defines the cutoff for rods, something like 5 or greater certainly *begins* to deserve the name long, thin rod. The math that defines its properties in gravitational and electromagnetic fields begins to differ significantly from that of globular particles and one is left with the question of how important would image analysis be compared to automated techniques that result in an equivalent spherical diameter (ESD).

The ESD concept is explored further in Appendix P2, "What is a Particle Size?" discussion. The rods shown here are helically-assembled, tobacco mosaic virus rods of length 275 nm and diameter 17 nm. So, the aspect ratio is about 16.

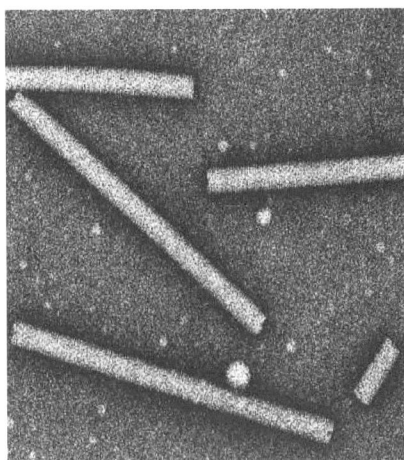

Non-spherical Particles, Discs or Platelets: An example from the inorganic world of clay platelets is shown here. These may have started out as hexagonal thin discs, but abrasion has taken its toll.

While this shape too has an ESD, the bigger the ratio of longest disc surface chord

to thickness, the more important shape analysis becomes.

Summary: Particle characterization covers a vast array of interesting specks of matter all the way from large molecules (globular proteins, polymers), to micelles, microemulsions, viruses, liposomes, latexes, pigments, clays, organic and inorganic oxides, sand, gravel, etc. By learning about particle characterization in several of these seemingly different fields, one can gain a better perspective about one's own field of interest.

Appendix P2: What Is Particle Size?[©]

By Bruce B. Weiner Ph.D., November 2010

Introduction: The short answer is this: diameter or radius. But what type of diameter or radius, and what to do about non-spherical particles? The following introduces some ideas for those who have never done particle sizing.

Length vs. Mass: If you are a particle technologist, then the only answer is length. But at a recent biochemistry national meeting, a group of protein chemists continually referred to the molecular weight of a globular protein, a relative molar mass, as the protein's "size". Over hearing this, our group was surprised because we refer to the size of a globular protein in nanometers. Apart from these protein chemists, the rest of us will mean a length in nanometers or microns when we refer to particle size.

To Be or Not To Be a Sphere: Of all the three-dimensional particles, the sphere is by far the most important in particle sizing. Why is that? Is it because most particles are spheres? No, though many come close to it (unaggregated latex, monoclonal antibodies, oil-in-water and water-in-oil emulsions, spherical micelles, liposomes, etc.). And still more are nearly so, especially if measurements are averaged over rotationally diffusing particles[i]. Over the timescales of many types of measurements, we are measuring a rotationally averaged size and thus a sphere represents, often, a reasonable approximation. In addition, if highly irregular particles are broken down due to abrasion, long ones are broken down into shorter ones and they become more globular rather than less. Think of wind and water action forming smooth, globular rocks from the jagged shards of volcanic debris. Think of irregular and/or jagged particles rounded off as they are mixed or stirred on their way to final product status.

According to the laws of thermodynamics, a liquid body with no external forces will form a sphere in order to minimize its surface area for a given volume of material. Thus, liquid droplets, ignoring external forces, form spheres. This explains why even cooling planets formed, to 1st order, spherical objects.

But perhaps the most important reason is that many 2nd-order differential equations that describe the physics of the automated methods used for measuring particle size are exactly soluble for spheres. Yes, we are fitting nature into what is conveniently achievable. The good news is that it works most of the time, most especially for quality control purposes.

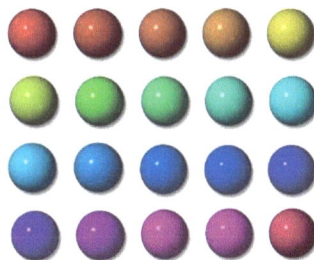

A Quick Tour of Spherical Geometry: The volume is either $4\pi r^3/3$ or $\pi d^3/6$ where twice the radius r equals the diameter d. If you can't easily recall these simple

147

formulas, consult an introductory math book as your first step in learning about particle sizing. For completion, the surface area of a sphere is either $4\pi r^2$ or πd^2. The Greeks knew these things 2,500 years ago and many of them still do.

The simple factor of two that relates radius to diameter is sometimes the cause for a 100% error. If a specification for size does not list it as radius or diameter, or if a result (mean size, for example) does not say which it is, or a graph is not labeled, then sometimes one is left to guess. A claim of being able to measure up to 5 micron in radius is the same as the claim of 10 micron in diameter. Look for this error. It occurs quite often and sometimes for purposes of deception, especially in advertising brochures.

The Equivalent Spherical Diameter, ESD: There are two types, geometric equivalent and measurement technique equivalent.

Geometric ESDs: Consider a static, 2-D image of a particle. First, it is two, not three dimensional. So, what looks like a circle might correspond to a thin disc-like particle and not a sphere, unless shadows reveal a more space-filling structure. When the image was taken, assuming it was not a 3-D holographic image, one assumes that as the particles settle, they do so, to become stable on a flat surface. Thus, most discs if they did not land on their faces would topple over if they landed on an edge. Try it. Throw coins into the air and see how they land.

There are many different geometrically defined ESDs that could be assigned to an irregular particle. One way to determine the ESD of an arbitrarily shaped particle is to draw circles around the actual image until it is just completely enclosed. Assign the diameter of the enclosing circle to that of the particle (d_e). Alternatively, find a circle whose area equals that of the measured particle area. This is easily done by counting pixels using computer programs that allow ever increasing accuracy with smaller and smaller pixels. Given the area of the drawn circle, assign its diameter to that of the particle ($\pi d^2/4$). This ESD should be labeled d_A. Or, one could trace the image's perimeter and assign that to the diameter of the circle with the same perimeter (πd_P).

These are all geometric ESDs. There are lots more choices based on parallel tangents (Feret's diameter, d_F) and chords (Martin's diameter, d_M). Note the fact that the same particle can have several different types of geometric ESDs, and if properly labeled, they should not be equal the more irregular the particle shape. And the ratio of two such geometric ESDs for the *same* particle says something about shape and space filling.

The more irregular the particle shape, the more difficult it is to describe it with just one parameter. Because of this, the interpretation of "size" determined by image analysis is more difficult than an automated machine based on an ESD determined by the technique. What is meant by this type of ESD?

Measurement Technique ESDs: Picture a stack of sieve plates. The mass of all the particles that remain on a plate (after suitable shaking) whose hole size[ii] is d_S are said to represent the cumulative increment by mass larger than diameter d_S. Think of a particle falling under gravity or moving radially outward in a centrifuge. Its velocity is measured and then set equal to that of a sphere that would have moved in the same way. The resulting diameter is called the Stokes' diameter, d_{St}, because the motion is described by Stokes' Law. Imagine a rotating, tumbling particle whose diffraction pattern is registered on a detector. Then the pattern is set equal to that of a sphere that would give the closest diffraction pattern. This is the so-called laser or laser diffraction particle size and should be labeled d_{LD}. Finally, consider using dynamic light scattering to determine the translational diffusion coefficient of a submicron particle. Then calculate the so-called hydrodynamic diameter or radius corresponding to the measured diffusion coefficient. There should be two subscripts here: H for hydrodynamic and DLS for the technique ($d_{H,DLS}$). In practice, double subscripts like this are rarely seen, *though if they were, it would become more obvious what was measured.*

In all these cases, and many more, the size of the sphere is assigned such that it would give the same result as the actual particle. These are so-called measurement technique ESDs (or ESRs). But what you see in practice is either d or r and this leads to confusion when comparing results. For spheres, if techniques were equally accurate, then a subscript would not be necessary. But for irregular shapes, using subscripts one would understand that the "sizes" should not be equal. And, as with geometric ESDs, ratios of measurement technique ESDs yield information on shape.

Unlike image analysis, there is only one definition for a given technique (ignoring specialized flow orientation techniques). Therein lays the weakness of image analysis: Which ones to choose to characterize the particle size? There are no generally applicable, easy guidelines to consult.

The Promise and Heart Break of Image Analysis: For very long rods, the aspect ratio, AR, is defined as the length divided by the diameter, L/d, and sometimes the reciprocal is called the AR. For more irregular particles, the longest dimension divided by the shortest dimension is the aspect ratio, or its reciprocal is. Given the L and d for each particle in a distribution, the aspect ratio can be calculated. A particle's performance could be correlated with L, d or perhaps with AR. There are only three choices here.

But imagine a more highly irregular particle, smooth or jagged. There are many possible statistical descriptors of shape and size. Most modern software offers dozens of choices that can be used. And that is the problem.

Which size parameter or subset of size parameters will correlate with particle performance? In some disciplines, answers are known. But in many they are not. Image analysis results in large amounts of data, but it does not necessarily result in immediately useful information.

There are other difficulties with image analysis such as depth of field, counting too few particles, and knowing when they are fused or separate particles when touching in a frame.

Three Types of Radii: First, there is the one we all picture, the hard-sphere, geometric radius, R_s. This radius is most easily obtained using image analysis. Second, as mentioned, in dynamic light scattering (DLS), we obtain the hydrodynamic radius, R_h. This radius is the one we get from comparing a sphere to the translational diffusion coefficient measured. Imagine a solid, hard-core particle whose surface is coated with long-chain polymers or surfactants that stick far out into the liquid. Sometimes called "hairy" particles, their radii are significantly larger than that of their cores. Finally, there is the radius of gyration, R_g, obtained from static light (SLS), small angle x-ray (SAXS), and small angle neutron (SANS) scattering. Interestingly, the R_g obtained from scattering measurements is independent of shape assumptions; whereas, any R_h value assume a sphere.

Ratios of R_g/R_h suggest shape: 0.77 a sphere; 1.54 a random-coil polymer.

Summary: After first determining that particle size is indeed a length and not the mass of a protein, determine if the results are given for a single statistical parameter or are multiple parameters involved using image analysis. If it is a single statistical parameter, is it a true diameter or an ESD determined geometrically (image analysis) or by comparison against what a sphere would yield using an automated technique (laser diffraction, centrifugation, sieving, zone counters, etc.). And, finally, is it a radius or a diameter (a true one or ESD/ESR) that is being discussed? With answers to these questions, you will be in a better position to compare numerical results more effectively. And that is the subject of the next application note in this series, Appendix P3.

----

[i] The rotational diffusion coefficient, D_R, for a sphere of radius 1 micron in water at 25 °C is 0.18 s^{-1} and it varies inversely with the cube of radius. Thus, a 100 nm radius particle is diffusing (rotating) 180 times per second. If the measurement time is a second or longer, the results are rotationally averaged.

[ii] Sieve sizes are a complete topic in themselves. Often, they are not circular holes. Abraded holes as well as particles broken by abrasion may be problems. Sifting long enough to ensure all smaller particles made it through is an issue. Finally, for highly irregular shapes, if the particle can be oriented by sifting, then it is the smaller dimension that is determined Think of a distribution of long rods of varying lengths

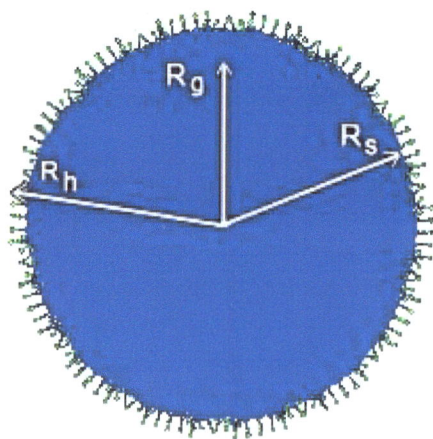

and varying, but much smaller diameters. Although unlikely, you would be determining the size distribution of the rod diameters and learn nothing about the distribution of rod lengths if you could sift them such that they all stood on end when passing through the sieves' holes.

Appendix P3: What is a Continuous Particle Size Distribution?[©]

By Bruce B. Weiner Ph.D., March 2011

Particle size distribution data can be presented numerically (tabular format) or graphically. When presented graphically, there are two types: differential and cumulative. They are related. If one differentiates the cumulative distribution curve, the differential distribution is obtained. If one integrates the differential distribution curve, the cumulative distribution is obtained.

Differential Distribution: The differential distribution shows the relative amount* at each size. For example, for the one shown here, using a ruler to draw horizontal and vertical lines, one can determine that the differential amount at 14.5 nm is about 40 while at 18 nm it is about 20. Therefore, the differential distribution is telling us there is twice the amount at 14.5 nm compared to the amount at 18 nm. In addition, the area under the curve between any two diameters divided by the total area under the curve equals the fraction of the total amount between those two diameters.

Measures of central tendency such as the modal and mean diameters are determined from the differential distribution. The modal diameter is the diameter at the peak of the differential distribution. In this example it is 8.5 nm. The mean diameter is the average diameter. In this example it is 10.7 nm.

This distribution is unimodal (single peaked) but not monodisperse (all one size). It has a width. There are several measures of width just as there are several measures of central tendency. One measure of width is FWHM, the

full width at half maximum. It is obtained by drawing a horizontal line at 50% of the maximum and taking the difference between the two places it intersects the distribution. In this example it is 8.4 nm.

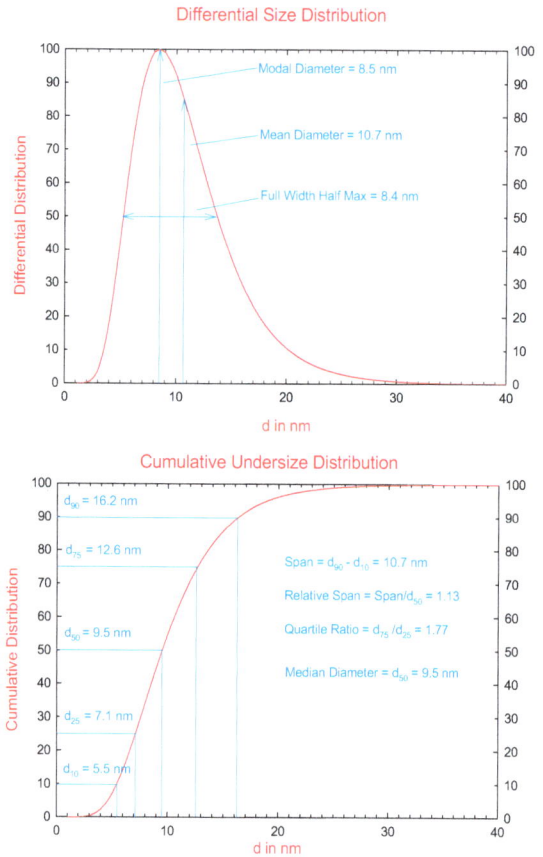

Differential Size Distribution

Cumulative Undersize Distribution

HWHM, the half width at half maximum, is another measure of width. It is defined as FWHM/2. In this example it is 4.2 nm.

FWHM and HWHM are measures of absolute width. Both have the same units as the diameter, nanometers in this example. A relative fractional measure of width is obtained by dividing either FWHM or

HWHM by the measure of central tendency from which it was derived, the modal diameter. In this example the HWHM/Modal Diameter is 4.2/8.5 = 0.49. A relative percent measure of width is then 49% in this example. Neither measure of relative width has units.

Given that a differential particle size distribution looks like the distribution of repeated measurements of any quantity, it is not surprising that the mathematics of probability distributions permeate the descriptions used in particle size distributions; namely, average (mean), variance, standard deviation, etc. Unfortunately, variance and standard deviation *also* suggest error or uncertainty in any measurement. Indeed, there is uncertainty in particle size distribution measurements; however, when the width of the differential size distribution is represented by the standard deviation of that distribution, it is not meant to suggest it represents the error in the measurement. Rather it is another way of representing the width of the size distribution. In fact, with repeated measurements of the size distribution, one could specify the standard deviation (the error) of the distribution's standard deviation (width of the distribution).

The standard deviation is not shown in the differential distribution above in order to keep the graph readable. If it was, one could define absolute and relative (fractional and percent) standard deviations (dividing by the mean diameter).

The differential distribution shown here is not symmetric about the modal diameter. If it was, and because the diameter axis is linear, then the modal, mean, and median

(defined for the cumulative distribution) diameters would all be equal. Such is not the case here. The distribution is skewed to larger sizes. There are various definitions of skew that are derived from probability distributions. In all cases, the skew is positive when the curve tails to the right more than to the left, and it is negative when the curve tails to the left more than to the right. The reference point for tailing is with respect to the modal diameter. A symmetric differential distribution has zero skew.

Cumulative Distribution: The corresponding cumulative distribution is also shown. The cumulative undersize distribution shows the relative amount* at or below a size. In this example 50% of the amount of particles is at or below 9.5 nm. Ninety percent are at or below 16.2 nm. These are just two of the possible percentile diameters.

In addition, the cumulative undersize distribution can be used to read the percent between any two diameters. For example, using the example shown above, if d_{90} = 16.2 nm and d_{75} = 12.6 nm, then it follows that 15% of the amount of particles have sizes between 12.6 and 16.2 nm.

The median diameter is another measure of central tendency. It is the diameter at the 50th percentile, designed d_{50}. Quartile diameters include d_{75}, d_{50}, and d_{25}.

There are several measures of absolute width one can derive given the cumulative distribution. One common measure is the span, $d_{90} - d_{10}$. In this example it is 10.7 nm. A dimensionless measure of width is the relative span defined as span/d_{50}. In this example it is 1.13. Other relative measures of width include percentile ratios such as d_{90}/d_{10} and d_{75}/d_{25}. In this example these values are 2.95 and 1.77, respectively.

The narrower a distribution is the more closely the absolute measures of width approach zero: FWHM, HWHM, variance, standard deviation (square root of variance), and span. Whereas, most of the relative measures of width like d_{90}/d_{10} and d_{75}/d_{25} approach unity.

The following table summarizes which type of distribution is used to determine various measures of central tendency and measures of the width of the distribution.

* The amount could be by number, by surface area, by volume, by mass, by intensity of scattered light. The distributions "weighted" by these quantities are all shifted with respect to one another. The topic of weighting is discussed in Appendix P5.

Measures of Central Tendency	Distribution Determined From
Mode	Differential
Mean	Differential
Median, d_{50}	Cumulative

Measures of Distribution Width	Distribution Determined From
Absolute	
FWHM	Differential
HWHM	Differential
Standard Deviation	Differential
Span, $(d_{90} - d_{10})$	Cumulative
Relative	
FWHM/Mode	Differential
HWHM/Mode	Differential
Standard Deviation/Mean	Differential
Quartile Ratio d_{75}/d_{25}	Cumulative
d_{90}/d_{10}	Cumulative
$(d_{90} - d_{10})/d_{50}$	Cumulative

Which Distribution Should I Use To Best Present My Data? It depends on what is customary in your field. For example, long ago tire manufactures correlated the relative strength of tire treads and tire walls to the quartile ratio d_{75}/d_{25}. So, to represent the distribution width use this relative width obtained from the cumulative distribution.

Appendix P4: What is a Discrete Particle Size Distribution?[©]

By Bruce B. Weiner Ph.D., May 2011

In the previous entry, Appendix P3, on particle size and size distribution, the main features of the differential and cumulative size distribution functions were defined and discussed using continuous distributions. Continuous distributions can be measured approximately in fractionation experiments[†] where, for example, data is taken every second as fractionated slices of the distribution pass by a detector yielding size and amount[*]. Nevertheless, particle sizes are discrete by nature. So, let us examine a discrete distribution, one determined by a single particle counter.[‡] Here the "amount" means the number or count or frequency of particles in a size class defined with upper and lower limits.

Cumulative Distribution Tabular Format:

Size Class Microns	Count	%	Cum %
0-4	104	10.4	10.4
4-6	160	16.0	26.4
6-8	161	16.1	42.5
8-9	75	7.5	50.0
9-10	67	6.7	56.7
10-14	186	18.6	75.3
14-16	61	6.1	81.4
16-20	79	7.9	89.3
20-35	103	10.3	99.6
35-50	4	0.4	100.0
>50	0	0.0	100.0
Totals	1000	100%	

Cumulative Distribution Tabular Format: There are 1,000 total particles counted (2nd column) and so it is easy to calculate the percent in each class (1st column) simply by moving the decimal point one place to the left (104 becomes 10.4%, 3rd column). And the cumulative percent undersize (4th column) is obtained by adding the current percent in size class to the total above it. Thus, the first one is 10.4%; the second is 10.4% + 16.0% = 26.4%; the third is 16.1% + 26.4% = 42.5%; and so on until one reaches 100%.

Cumulative Distribution Graphical Format: If we plot the cumulative percent vs the upper limit of each size class, we obtain the cumulative undersize distribution. See plot above.

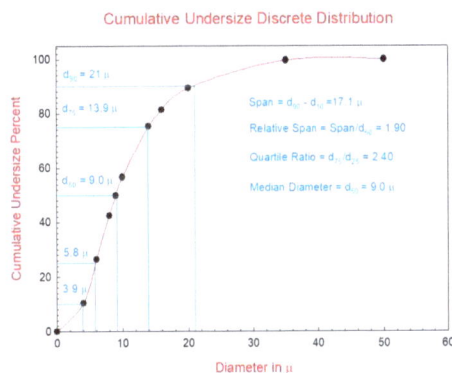

Cumulative Undersize Discrete Distribution

The example here is made up of 10 size classes of unequal size, which has consequences when the frequency distribution is discussed. But first let us discuss the cumulative undersize distribution as it is the more straightforward when there are relatively few size classes.

155

The percentile diameters were read from the graph; however, in this example, they could have been estimated directly from the tabular data with the median diameter, d_{50}, falling exactly at 9 μ.

The black points are the raw data with one added at zero to guide the eye. The red line is a smoothed version of what the continuous cumulative undersize distribution might look like. Obviously, given so few data points, several other smoothed curves could have been drawn. Or, if you assumed you knew the simple functional form of the cumulative undersize distribution, you could have done a nonlinear least squares fit to it. Problem: Most size distributions don't fit simple functional forms. Occasionally some do, like a Lognormal (two parameters), but not many.

If you don't have a simple functional form that you can differentiate, or you have too few data points such as is the case here, then numerical differentiation is extremely suspect. In that case, how do you obtain the equivalent of the differential size distribution in order to determine mean, mode, and measures of width? Answer: Start with a histogram.

Histogram of the Frequency Distribution: The table shows the count (the number) of particles (2nd column) vs size class (1st column). In statistics this is called the frequency of occurrence in each size class. This is plotted in the lower left. But it doesn't look much like a size distribution, does it? The reason is simple: The size classes are not all equal and so until we do

something to fix that, we can't discern anything analogous to the differential size distribution shown in the preceding appendix.

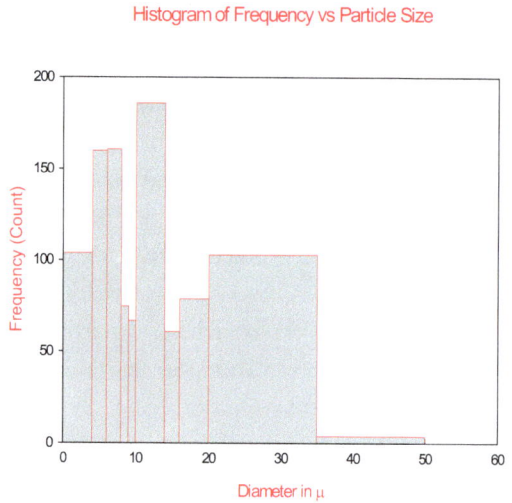

Histogram of Frequency vs Particle Size

To remedy the situation, divide the count by the size class width. This yields the Frequency/micron (3rd column in table below). This allows us to compare distributions made with different size classes.

Size Class in Microns	Count	Freq/	Frac/μ
0-4	104	26.0	0.0260
4-6	160	80.0	0.0800
6-8	161	80.5	0.0805
8-9	75	75.0	0.075
9-10	67	67.0	0.0670
10-14	186	46.5	0.0465
14-16	61	30.5	0.0305
16-20	79	19.8	0.0198
20-35	103	6.87	0.0687
35-50	4	0.267	0.0003
>50	0	0	0
Total	1000		

Plot of Frequency/μ vs. Size: A plot of the 3rd column vs size class (1st column) begins to look like a size distribution:

Histogram of Frequency/μ vs Particle Size

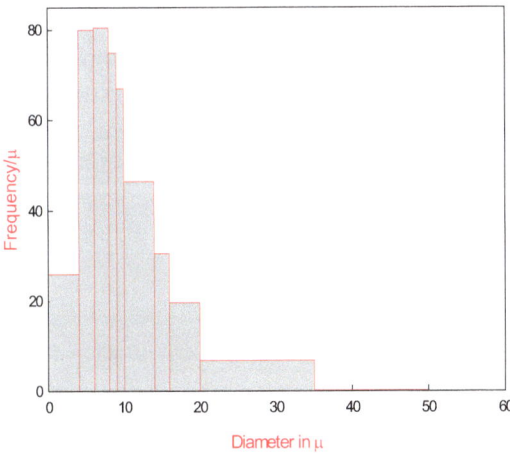

You can see a modal diameter somewhere near 6 μ. And you can calculate a mean diameter as follows: $\overline{d_n} = \dfrac{\sum_i N_i \cdot d_i}{\sum_i N_i}$ where

N_i is the value in the 2nd column, the frequency (count or number) of particles, and

d_i is the midpoint of the class size from the 1st column. Sum over i size classes from 1 to 10, and the result is the number or count mean diameter of 10.8 μ.

You can even estimate the full width at half maximum: Since 80 is approximately the maximum value, 40 is half. Draw a horizontal line at 40. It crosses the unimodal plot at 4 μ and 14 μ. Thus, the FWHM = 10 μ and the HWHM = 5 μ.

Now the area in any bar represents the total amount, in this case number of particles, in the size class. This is analogous to the area under the continuous differential distribution curve between any two diameters. With a frequency distribution, it was the height of the bar that represents the total number in class.

Therefore, with the Frequency/μ plot representing the equivalent of the differential size distribution, we can again determine mean, mode, FWHM and other measures of size distribution width, absolute and relative. However, here is what we can't do: We can't compare this distribution to another one unless the total number of particles counted was equal.

Plot of Fraction/μ vs Size: In order to compare distributions with different total counts, we must divide each Frequency/μ by the total number of particles (1,000). This produces the Fraction/μ, the 4th column. A vertical bar chart of Fraction/μ vs. size class is shown on the next page:

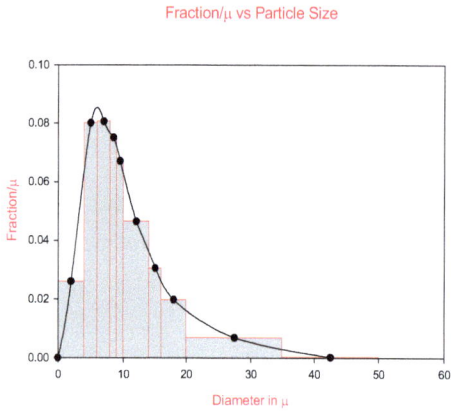

Fraction/μ vs Particle Size

Its shape is the same as that of the Frequency/μ; however, the y-axis is now fraction/ μ. If you made measurements with different amounts of sample, you could overlay them in this type of plot to see if they were equal or not.

One more thing can be done here to show how a discrete Fraction/(Size Class Width) plot begins to resemble a continuous differential distribution. Connect the midpoints of each bar using a smooth, continuous curve. This results in something that looks like the differential size distribution. Though there are many such smooth curves that could be drawn.

Which Discrete Distribution Should I Use To Best Present My Data? If there are relatively few size classes and no reason to believe a function fits it well, then only use statistics obtained from the cumulative undersize distribution: median, percentiles, span, etc. These numbers will have the least error.

Summary: Like the continuous distribution case, one can construct a cumulative undersize discrete distribution from discrete tabular data and determine median, percentiles, and measures of distribution width. It is more difficult to construct

something akin to the continuous differential distribution.

However, to do so, start with the Frequency vs Size Class table and plot. If all the size classes are equal, then divide by the total amount of particles and create a Fraction/(Size Class) plot.

If the size classes are not equal, divide each frequency by its own size class width and then divide by the total amount.

If the number of size classes is adequately large, see if a smooth differential-like distribution can be constructed by connecting the midpoints.

† Sedimentation, both gravitational and centrifugal, field-flow fractionation, hydrodynamic chromatography, etc. are examples.

* The amount could be by number, by surface area, by volume, by mass, by intensity of scattered light. The distributions "weighted" by these quantities are all shifted with respect to one another. The topic of weighting is discussed in Appendix P5.

‡Single particle counters include electro- and optical-zone counters, image analyzers, and single particle tracking devices.

Appendix P5: What Is Particle Size Distribution Weighting: How to Get Fooled about What Was Measured and What it Means?[©]

By Bruce B. Weiner Ph.D., June 2011

Introduction: Some particle size instruments determine size for individual particles. They are single particle counters. Some instruments determine surface area as a function of particle size. Some instruments determine mass or volume vs size. And some instruments determine various functions of scattered light intensity as a function of size. All can produce particle size distributions. And, in principle, one can transform from one type to the other in order to compare results. For example, if a measurement with a single particle counter produced a differential number-weighted size distribution and you wanted to compare the results to a measurement with another type of instrument that produced a differential volume-weighted size distribution, what would you do?

Transform One Weighting To Another: You could convert the differential number-weighted distribution to a differential volume-weighted distribution or the other way around. How do you do that? It's easy. For each size class in the *discrete* Number Fraction/(Size Class) distribution[*], multiply the number or count by the diameter cubed (spheres are assumed). In the case of discrete distributions, the diameter to be cubed is most likely the midpoint of the size class. The result is a *discrete* Volume Fraction/(Size Class) distribution. And to convert a *continuous*

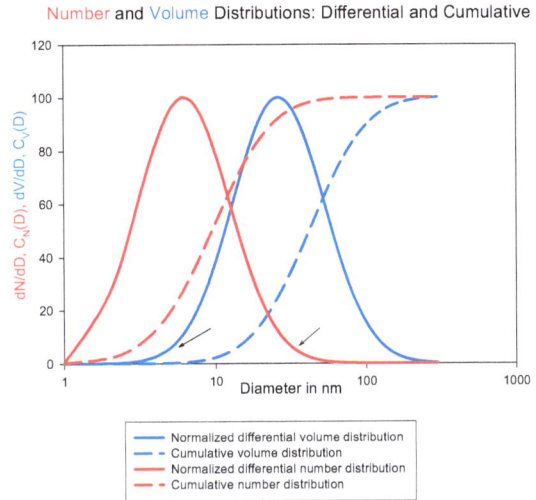

Number and Volume Distributions: Differential and Cumulative

differential number-weighted distribution into a *continuous* differential volume-weighted distribution, multiply by D^3. The result is a *continuous* volume-weighted differential distribution. Of course, the initial numerical values will not be normalized, but that is easy to remedy. Here is a simple example using *continuous* distributions:

The bell-shape curves are the differential distributions weighted by number, dN/dD, or by volume, dV/dD. The sigmoidal curves are the cumulative distributions weighted by number, $C_N(D)$, or by volume, $C_V(D)$.

Perhaps the initial measurement determined dV/dD, from which by integration $C_V(D)$ was determined. Then, to determine the unnormalized dN/dD, use the following formula:

$\dfrac{dV}{dD} = \dfrac{dN}{dD} D^3$. At each y-value of the differential volume-weighted distribution, divide by D^3. Find the maximum value in the resulting set of unnormalized numbers. Then determine the factor that will make that value 100. Apply the same factor to all the other unnormalized values. Now the differential number distribution is normalized.

Relationship Amongst Transformed Distributions: Notice that dV/dD is always shifted to the right, towards larger D's, than dN/dD. Likewise, the cumulative distributions are shifted in a similar fashion. The modal diameter—corresponding to the peak in the differential distribution—and the median diameter—equal to the diameter at 50% of the cumulative distribution--of the volume distribution is always higher than the corresponding ones in the number distribution.

If these were nonporous particles**, then the surface area-weighted differential distribution, dS/dD, is related to the number-weighted by the following simple equation:

$\dfrac{dS}{dD} = \dfrac{dN}{dD} D^2$. And the curves when normalized and plotted would sit between the number- and volume-weighted for both the differential and cumulative representations.

First Warning: Which Weighting Was It? If someone says the median diameter is 10.0 nm and it's a broad distribution, if you don't ask what the weighting was, you are missing a lot of information. In the graph

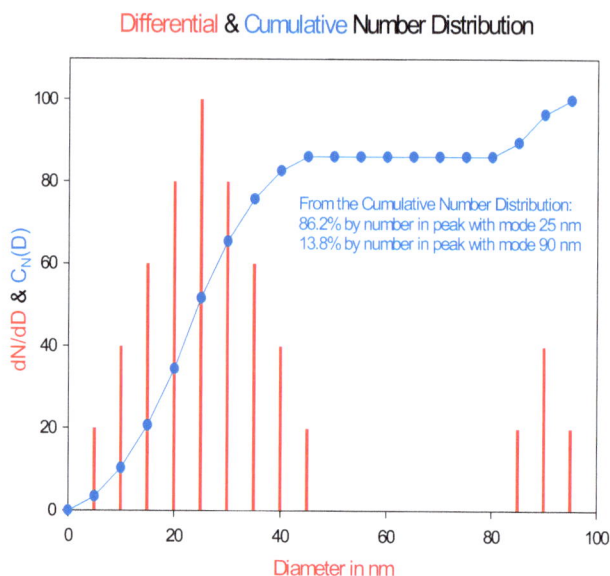

Differential & Cumulative Number Distribution

From the Cumulative Number Distribution:
86.2% by number in peak with mode 25 nm
13.8% by number in peak with mode 90 nm

above, the number median diameter is 10.0 nm and it is a broad distribution; yet, the volume median diameter is 42.3 nm and a large fraction by volume of the particles are well above 10.0 nm.

Second Warning: Transformations Can Be Dangerous: Look again at the graph. Notice the two arrows, one at the base of each differential distribution. A relatively small volume of particles in the left-hand tail of dV/dD is responsible for a large portion of the dN/dD distribution; likewise, a relatively small number of large particles in the right-hand tail of dN/dD are responsible for a large portion of the dV/dD distribution. This situation is often a prelude to disaster.

The reader probably assumed that all the particles were used in calculating the results as shown. Indeed, the author did exactly that. But in real measurements, this is usually not the case: the information in the tails of the differential distribution is often not accurately known.

In counting experiments, one tends to under count a relatively few large particles. But these are the very ones that dominate the volume distribution. Therefore, the calculated volume distribution is shifted much too low. And in experiments that use scattered or diffracted light, the small particles don't contribute much to the signal. Thus, they are underrepresented in the distribution with the most natural weighting (intensity-weighted in this case). Therefore, the calculated number distribution is often shifted much too high. Transformations, though simple algebraically, may be very inaccurate for these reasons.

Since the 90 nm peak was narrow, it is not surprising that it remains the mode. Whereas, the lower peak is somewhat broad and just like the case of the broad unimodal distribution examined earlier, the entire peak shifts to the right.

Finally, the volume-weighted distribution continues the trend: The larger peak centered on 90 nm contains, by amount, most of the volume; whereas, it contains the least by number.

Notice the cumulative distributions have a plateau. This is characteristic of multi-modal distributions. Either from a corresponding tabular presentation or from the graph, you can read off the amount in each peak by finding the plateau value.

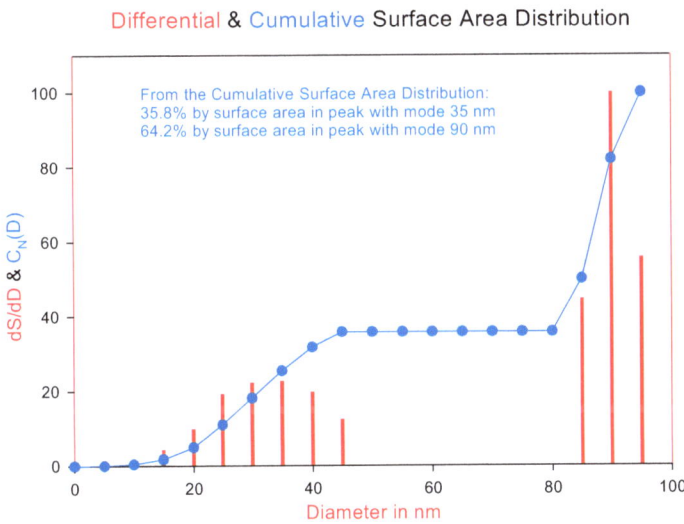

Differential & Cumulative Surface Area Distribution

From the Cumulative Surface Area Distribution:
35.8% by surface area in peak with mode 35 nm
64.2% by surface area in peak with mode 90 nm

Which Weighting Should I Use To Best Present My Data? Sometimes the answer is dictated by the field of study you are in: When measuring blood, sperm, or micro-contaminates, the absolute number per unit volume is required. Use a single particle counter and stay with the number distribution. If the particles are used by mass, use the volume distribution (here the assumption is all the particles have the same density; if not the volume and mass distributions are no longer equal).

Another Example Using a Bimodal Distribution: The peak centered on 90 nm has a much smaller relative number of particles than the peak centered on 25 nm. In addition, the 90 nm peak is narrower. When the transformation is made to surface area-weighted, it should not be surprising that the amount by surface area has shifted to the larger sizes: 35.8% by surface area centered on 35 nm mode and 64.2% by surface area centered on 90 nm mode.

Differential & Cumulative Volume Distribution

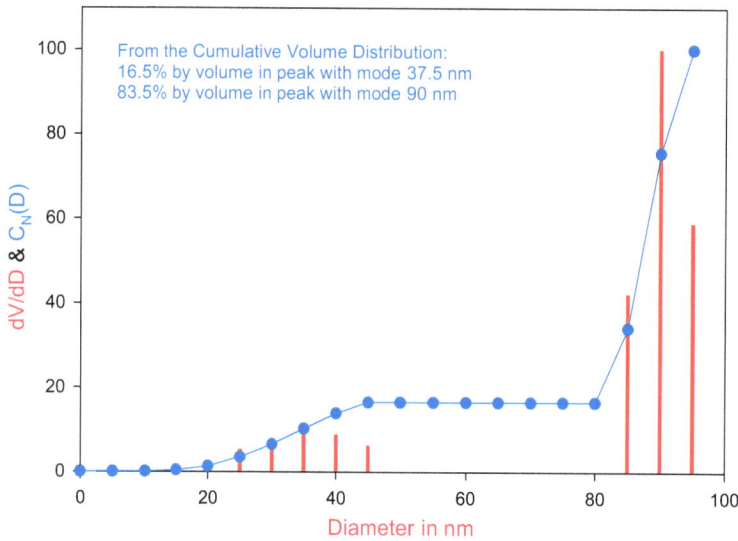

From the Cumulative Volume Distribution:
16.5% by volume in peak with mode 37.5 nm
83.5% by volume in peak with mode 90 nm

relatively small fraction of large particles that carry most of the volume and mass of the distribution while counting particles; and missing a significant but relatively large number of small particle that carry most of the number-weighted information, because they don't contribute much to the intensity of scattered light.

*For a definition of Fraction/(Size Class), see Appendix P4.

**Porous particles that have a lot more surface area. Surface-area weighted size distributions should not be calculated from either number- or volume-weighted ones unless it is certain the porosity is unimportant. This is the case for liquid droplet particles but not the case for many oxide particles.

Summary: As important as it is to know if the "size" is a true, spherical diameter or radius, or an equivalent spherical size determined by the measurement technique, it is equally important to know if the distribution is weighted by number, surface area, volume-mass, or intensity. Without this information, you really don't know what emphasis to put on the results.

It can't be emphasized enough that while the algebra for transforming one distribution to another is simple enough, it is the assumption that all the particles have been measured that is usually wanting. Common errors include missing a significant but

Appendix SLS1: Differential Refractive Index Increment, dn/dc

Definition of dn/dc

The differential refractive index increment, sometimes referred to as the specific refractive index increment, or SRII, is the slope of n_s vs. concentration at zero concentration. Here n_s is the refractive index of a protein or polymer solution. The SRII is a limiting slope at zero concentration often referred to as just "d n d c". However, in most solute/solvent samples n_s vs. concentration is a straight line and the limiting slope is the same as the slope at finite concentrations, even up to 10's of mg/mL.

Defining dn/dc

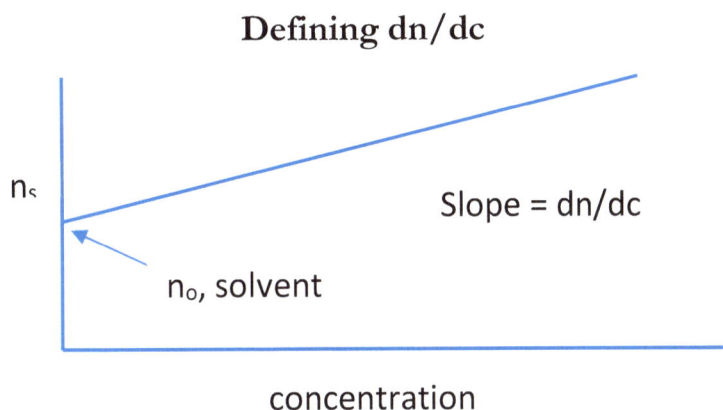

These concepts are illustrated above and serve to define conceptually dn/dc. In practice, the variation in n_s with temperature is larger than its variation with concentration, and the above plot is nearly impossible to obtain with precision unless extreme measures are taken for temperature stability. To avoid this problem, one should use a differential refractometer, not a classic Abbe refractometer.

Differential Refractometer

In a differential refractometer, the solution and solvent are maintained at the same temperature in the same cell separated by a transparent partition. It is not important to know the exact temperature to within a few degrees; it is only important that the temperature difference between the two sides of the cell is small, $\leq 0.01\ °C$. This is not a particularly stringent condition. It can be accomplished in a well-insulated, small cell by waiting for thermal equilibrium or by controlling the cell temperature to within $\pm\ 0.01\ °C$.

The signal in a differential refractometer is proportional to the displacement of a beam of light as it traverses the solution and solvents sides of the cell. To improve resolution, the beam is reflected through the cell to double the displacement. The displacement is proportional to the difference in refractive index of the solution and solvent-sides of the cell. $\Delta n \equiv n_s - n_o$.

The value of dn/dc normally varies from about 0.05 to 0.25 mL/g. Occasionally, negative values occur when the refractive index of the solvent is greater than that of the polymer molecule. Since dn/dc is squared in the equations used to determine M_w, a negative dn/dc is physically acceptable.

Establishing Error Limits

In batch mode SLS measurements, $M_w \propto (dn/dc)^2$. Therefore, the relative error in calculating M_w is twice the relative error in dn/dc. For example, if you want to know M_w to 5% using static light scattering, a reasonable error limit on M_w, then the error contributed by dn/dc must be less than 2.5%.

A special case arises when using the *same* differential refractometer in a GPC/SEC system to determine both concentration and dn/dc. In this case, $M_w \propto (dn/dc)$, and the relative error in calculating M_w is equal to that of dn/dc.

Literature Values of dn/dc

Literature values for a very large number of common polymers in a variety of solvents are tabulated in:

Huglin, M.B., "Specific Refractive Index Increments", Chapter 6 in Light Scattering from Polymer Solutions, M.B. Huglin editor, Academic Press, London & New York, 1972.

Huglin, M.B., "Specific Refractive Index Increments of Polymers in Dilute Solution", pp. IV-267 to IV-308 in Polymer Handbook, 2nd ed., J. Brandrup and E.H. Immergut, editors, Wiley-Interscience, NY, 1975. Essentially a repeat of values tabulated in reference 1.

Timasheff, S.N., p. 372-382 in the Handbook of Biochemistry and Molecular Biology, 3rd edition, Vol. II, edited by G.R. Fasman, 1976. Values for proteins in water.

Refractive Increment Data-Book for Polymer and Biomolecular Scientists, compiled by A. Theisen, C. Johann, M.P. Deacon, and S.E. Harding, Nottingham University Press, 2000. Many entries are more recent than the classic compilations above.

The classic values (before the widespread use of lasers, say before 1970) are given primarily at 436 nm and 546 nm, the mercury lines. Hg-arc sources were the light sources used in differential refractometers and light scattering instruments before lasers and LEDs became widely available. As shown below (Wave Length Dependence of dn/dc), fitting to $A + B/\lambda^2$ using values known at two wavelengths is reasonable for calculating at laser wavelengths of 488 nm and 514.5 nm (Argon-ion laser lines) or 532 nm (newer, frequency-doubled, solid state lasers). However, using the fit to predict dn/dc at 632.8 nm (HeNe laser line) or 660 nm ("red", diode lasers) is less acceptable, but may be necessary.

Without a direct measurement, literature values may be all that you have to work with. It is important, however, to understand that dn/dc varies with the solvent's refractive index, impurities in the solvent and polymer, molecular weight of polymer, temperature, and wavelength. Each of these variables is discussed below.

Effect of Solvent's Refractive Index

If a polymer can be dissolved in several different solvents, it is best to choose the one with the greatest difference in refractive index. In this case, dn/dc will be a maximum (scattering intensity varies as square of dn/dc), and errors in determining dn/dc will become less important. While the shape of the molecule (affecting R_g) and its interaction with the solvent (affecting A_2) may change from solvent-to-solvent, M_w will not.

The variation of dn/dc with solvent refractive index is roughly linear. Here is an example.The data is from Huglin's book, reference 1 above. Note the approximate factor of two in the spread of dn/dc. This leads to a factor of four in the scattered intensity. So, using acetone instead of chloroform will yield greater excess intensity, something easier to measure with greater confidence. Note that the near linearity for this polymer in several solvents suggests that the partial molar volume at infinite dilution is the same in all the solvents.

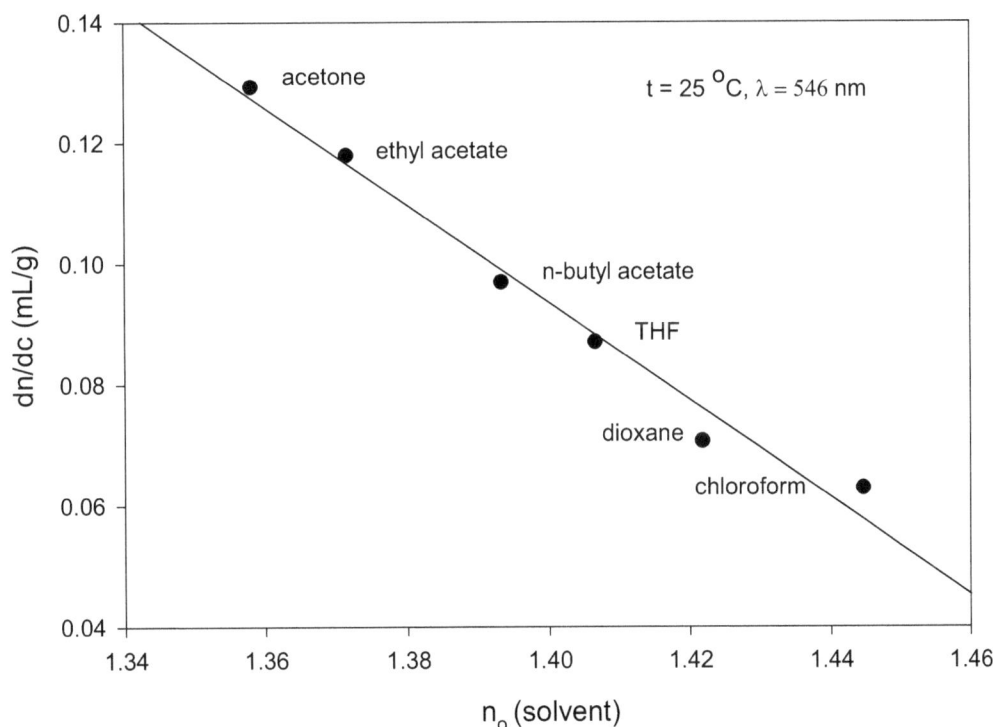

dn/dc of PMMA in Different Solvents

Effect of Impurities in Solvent and Polymer

Literature values from several authors for apparently the same polymer/solvent system often differ by several percent, even at the same temperature and wavelength. While this can be due to sample preparation, calibration, and measurement errors, it is sometimes due to impurities in the polymer or solvent. Solvent impurities will cancel when making the dn/dc measurements since they appear in both the pure solvent and the solvent used to prepare the solutions, but then your light scattering measurements must be made with the same polymer/solvent impurities as that used by the author of the literature article. As this is highly unlikely, it is better to measure dn/dc using your polymer/solvent.

Another manifestation of this same problem may occur if the solutions and solvent have not been kept under the same conditions for both the dn/dc and light scattering measurements. Ideally, the exact same solutions should be used for both measurements, and the solvent used to prepare the solutions should be kept under the same conditions (sealed, no additional moisture or other impurities allowed in contact). Otherwise, the subtraction involved in determining the excess scattered intensity and that involved in determining Δn may be different.

Molecular Weight Dependence of dn/dc

Generally, dn/dc increases with molecular weight and reaches a plateau when the end groups are sufficiently far apart. Below, approximately, 1,000 g/mol, dn/dc varies considerably. It increases by a few percent up to ~10,000 to ~ 20,000 g/mol, depending on the solvent, polymer shape and end-group contributions to the refractive index. Above this range it is, typically, constant.

An early example from the literature involved polyethylene glycol in water. Here are the results:

Variation of dn/dc with M for PEG/Water

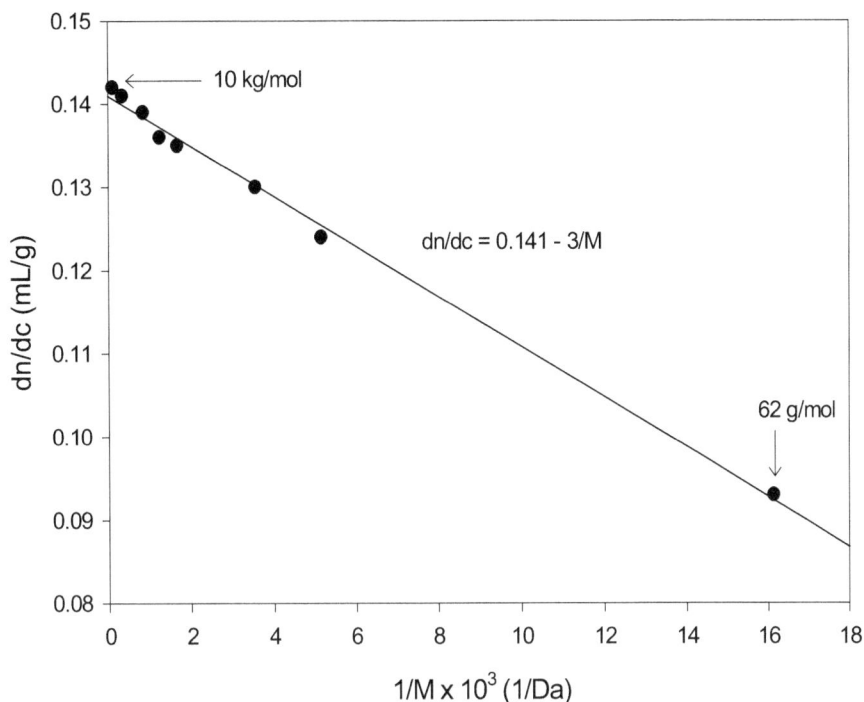

Generally, data like this fits reasonably well to $dn/dc = \alpha - \beta/M$, so when M is large enough, dn/dc does not change. Given the fit above, the change in dn/dc is less than 2% when M > 1,064 g/mol. This is an unusual case and may be because both solvent and polymer have hydroxyl groups.

Polystyrene in cyclohexane is a more common case as shown in the next graph. Here the rate of change with M is 9 times as large and the change in dn/dc is less than 2% when M > 7,750 g/mol. Both sets of data are tabulated in Chapter 6 of Huglin's 1972 book and in the original literature cited therein.

It is particularly important to realize that each set of data plotted here was measured by the same authors (Kratohvil, J.P. 1968 for PS/cyclohexane and Rempp 1957 for PEG/H2O). Therefore, it is safe to assume that the samples and corresponding solvents in each run were prepared under similar conditions for the different molecular weights.

If one simply compared a list of dn/dc values for the same solvent/polymer from different authors, all at the same wavelength and temperature, one would be surprised to find

Variation of dn/dc with M for PS/Cyclohexane

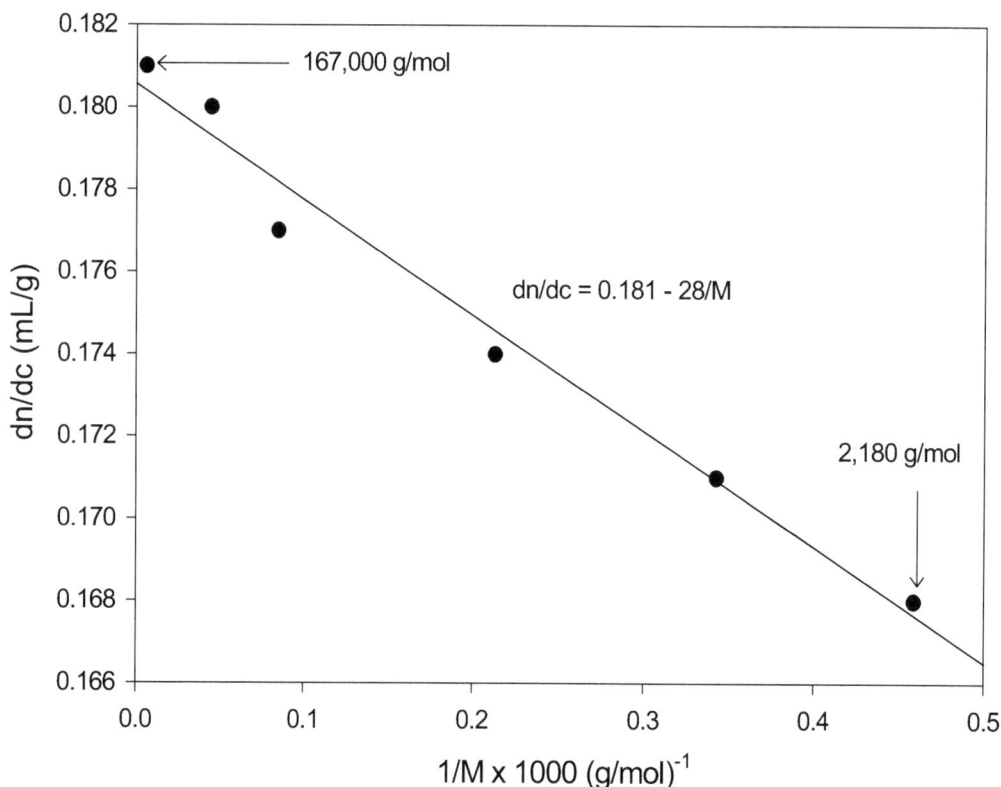

variations of several percent. While it is tempting to think that this variation can be explained by differences in molecular weight, more often the reason is that the impurity levels in the polymers and solvents are different from author-to-author. Therefore, if you want to deduce the molecular weight dependence, one must rely on data from the same author.

Temperature Dependence of dn/dc

The refractive index of liquids varies with temperature. For common organic solvents, the range is from -3×10^{-4} K^{-1} to -5×10^{-4} K^{-1}. Water's value, an exception, is -1.1×10^{-4} K^{-1}. Experimental temperature coefficients for dn/dc are usually linear from room temperature up to 100 °C and higher. The coefficients can be zero, negative, or positive, but the variation in the absolute value is typically in the range of 1 to 5×10^{-4} $mL \cdot g^{-1} \cdot K^{-1}$. Thus, a few degrees difference in the absolute temperature between the measurement of dn/dc and the light scattering experiment is not significant.

Wavelength Dependence of dn/dc

Experimental observations show that dn/dc decreases with increasing wavelength, with variations of 1 to 3% over the range of 436 nm to 546 nm. These two wavelengths, from an Hg-arc source, represent the two most common ones used in SLS measurements prior to the use of lasers. Since they span the 488 nm and 514.5 nm lines of an Argon-ion laser, and the 532 nm line of a frequency-doubled, solid state laser, one could interpolate (not linear interpolation, see equation below) values of dn/dc from those measured at the classical values. More uncertainty would accrue to values extrapolated to 632.8 nm, the wavelength of a HeNe laser, or to values extrapolated to ~670 nm, a common wavelength for a "red" diode laser. (Again, the extrapolation is not linear. See below.)

Except for absorption peaks in the spectrum of either the solvent or the polymer, the variation in wavelength is given by:

$$dn/dc = A + B/\lambda^2$$

This equation is also applicable to the variation in refractive index of solvent and polymer (again, assuming no absorption in the wavelength range of interest) and is known as the Cauchy equation. If dn/dc is known at two wavelengths, this equation can be used to calculate it at a third. However, it is much better if three or more measured values fit well to a straight line.

Some investigators ignore the wavelength dependence, and this is usually a mistake. Although there are some polymer/solvent pairs that exhibit very little dispersion, there are others that exhibit quite a bit. For example, poly(vinyl acetate)/water has an 8.8% variation in dn/dc from 436 nm to 632.8 nm (see Page 40 of reference 4 listed above). In batch mode, such an error would result in nearly an 18% error in M_w; whereas, PVC/dioxane shows no change in dn/dc from 436 nm to 586 nm (see page 40 of reference 4 listed above). In addition, many researchers working with proteins and protein-like structures assume a constant dn/dc of approximately 0.18, ignoring wavelength and temperature corrections.

Since the errors in dn/dc arising from differences in samples, even samples supposedly of the same chemical composition and with the same solvent, can be larger than the variations with temperature and wavelength, in order to find suitable A & B coefficients for the equation above, one must choose examples from the same author. The hope is that at least the polymer/solvents were the same. Here is a selection obtained from references 1 and 4 above.

Polymer/solvent	A	B
Polystyrene/toluene	+ 0.1015	+ 0.00200
Polystyrene/DMF	+ 0.1450	+ 0.00500
Polystyrene/MEK	+ 0.1963	+ 0.00675
Poly(methyl methacrylate)/Dioxane	+ 0.0486	+ 0.00031
Sucrose/water	+ 0.1392	+ 0.00115
Myosin/water	+ 0.1847	+ 0.00121
BSA/0.1M NaCl	+ 0.1791	+ 0.00378
Poly(acrylamide)/acetic acid	+ 0.1857	+ 0.00253
Poly(dimethyl siloxane)/toluene	- 0.0767	- 0.00504

The average value of B for all eight of the positive values is + 0.0028; for the non-aqueous samples the average is a bit higher at + 0.0035; and for the aqueous-based samples the average value of B is a bit lower at + 0.0022. The table is obviously not exhaustive, but the values do cover a range of polymer/solvents. B is calculated with λ in microns.

Here are examples of using the table to estimate small, first-order corrections.

Suppose you measure dn/dc = 0.1500 mL/g for a random-coil polymer in an organic solvent using the 620 nm source in a commercially available differential refractometer. You want to estimate dn/dc at 660 nm, the wavelength of the laser in your light scattering device. Assuming B = + 0.0035, calculate A = 0.1500 – 0.0035/0.6202. The result is A = 0.1409. Now calculate dn/dc at 660 nm. The result is dn/dc = 0.1409 + 0.0035/0.6602 = 0.1489 mL/g. The difference in dn/dc at 620 nm and 660 nm is 0.74%. The error in a calculated value of M_w using 0.1500 mL/g instead of 0.1489 mL/g is 1.5%, nearly insignificant compared to other errors. This case demonstrates why using the dn/dc values measured close to the laser wavelength is enough for many purposes.

Suppose you find a literature value of 0.1500 mL/g measured at 436 nm and you want to estimate the value at 632 nm. Aside from the fact that the literature value may correspond to a polymer/solvent system with different impurities or molecular weights than yours, this large wavelength difference cannot be ignored. In this case $A = 0.1500 - 0.0035/0.4362 = 0.1316$, and the estimated value at 632 nm is $dn/dc = 0.1316 + 0.0035/0.6322 = 0.1404$ mL/g. Now the difference between the two dn/dc values is 6.4%. This could lead to a 13% error in M_w in batch mode, something that may not be acceptable.

Included in the table are two values that demonstrate atypical situations. First, note the negative value of A and B for poly (dimethyl siloxane) in toluene. This means that dn/dc is also negative. Negative values of dn/dc arise when the refractive index of the solvent is larger than that of the polymer. Since, typically, this is not the case, most dn/dc values are positive. Second, notice the rather small value of A and B for PMMA in dioxane. This means that dn/dc is also small, and it arises because the refractive indices of polymer and solvent are similar. It also demonstrates the fact that there exist polymer/solvent combinations with near-zero dn/dc values. In these cases, the errors in light scattering measurements will be very large, and it may not be worthwhile using SLS at all.

Appendix SLS2: Rayleigh Ratios

In static light scattering one measures scattered intensities (more formally the radiance). But that depends on the incident irradiance, I_{inc}, from the light source, the observed and illuminated scattering volume, V_{obs}, and the distance from the scattering volume to the detector, r. And to calculate molecular and particle properties independent of these source and detector optical properties, one can define the Rayleigh ratio at scattering angle θ as:

$$R_\theta = I_\theta \cdot r^2 / (I_{inc}V_{obs}) \qquad \text{(SLS2-1)}$$

Since light is scattered not only by the molecule or particle of interest but also the solvent or dispersing liquid, it is only when the contributions from these entities are subtracted from the total scattering that the excess intensity of light scattered leads to the excess Rayleigh ratio which is related to molecular and particle properties of interest.

Define <u>excess</u> quantities as $\Delta I_\theta = I_\theta(\text{solution}) - I_\theta(\text{solvent})$ and $\Delta R_\theta = R_\theta(\text{solution}) - R_\theta(\text{solvent})$. Thus:

$$\Delta R_\theta = \Delta I_\theta \cdot r^2 / (I_{inc}V_{obs}) \qquad \text{(SLS2-2)}$$

Rayleigh ratios are not really ratios. Ratios don't have units. Rayleigh ratios are really values with cgs units of $cm^{-1} \cdot steridian^{-1}$. They represent the fraction of light scattered per unit length per unit solid angle. Given the long tradition of calling them a ratio, we continue with this traditional misnomer.

Calculation of V_{obs} and r can be complicated. And determination of I_{inc} requires an absolute photometer. For these reasons, a few absolute R values have been determined for pure liquids. Measurement of I_{90} in a light scattering device leads to an instrument calibration constant that is then applied to the measurement of the intensity from the unknown at different angles. Corrections are made for scattering volume differences at different angles [$\sin(\theta)$]; refraction corrections due to index differences between the <u>s</u>olvent and the <u>c</u>alibration liquid [$(n_s/n_c)^2$]; and reflection corrections [$(I_\theta - f_s I_{180-\theta})/(1 - f_s^2)$], where f_s is the reflection coefficient. The air-glass interface yields the largest f_s followed by f_s for water-glass. Index matching is a great help in this regard to remove the air-glass interface surrounding the sample cell.

The Rayleigh ratio of a pure liquid is a strong function of the wavelength, and it is best to use values measured at the wavelength of interest. In laser light scattering this is shown in Table SLS1.

Argon-ion Laser	Argon-ion Laser	Frequency Doubled Solid State Laser	HeNe Laser	Solid State Red Diode Lasers
488.0 blue	514.5 green	532 green	632.8 orange	675, 640, 637

Table SLS1: Common lasers and their wavelengths used in laser light scattering.

An instrument designed by Kaye[48] for absolute measurements of Rayleigh ratio (so $r^2/I_{inc}V_{obs}$ determined) using a vertically polarized HeNe laser gave the following results in Table SLS2-2 including column 4, the depolarization ratio using vertically polarized laser but no analyzer in the detector (ρ_U):

632 nm	$R_V = R_C(90°)$	ρ_V	ρ_U
Toluene	14.0×10^{-6} cm^{-1}	0.325	0.491
Benzene	12.6×10^{-6}	0.265	0.419

Table SLS2-2: Absolute Rayleigh Ratios at 632.8 nm.

Recall from Chapter II, the relationship between the depolarization ratio using a polarizer on the detector and one using no polarizer:

$$\rho_U = 2\rho_V/(1 + \rho_V) \qquad \text{(SLS2-3)}$$

Kaye measured R_V and ρ_V, which was used to calculate ρ_U. The absolute errors estimated for R_V are +/- 2%. Confirmation came a decade later when Cannell[49] and coworkers designed a very sensitive but not absolute light scattering instrument. They also used a vertically polarized, HeNe laser. Calibrating using Kaye's value for toluene, the Rayleigh Ratio for two other liquids was found to agree with Kaye's values for those liquids. Such excellent agreement between two completely different types of light scattering devices over a 10-year period strongly suggests the values in the table above are very reliable.

Before lasers, two wavelengths were isolated from Hg-arc sources and used in light scattering: 546 nm and 436 nm. R_U and ρ_U were measured and R_V and ρ_V were calculated. Coumou's data[50] on several liquids are considered very self-consistent, and they are often referenced for use in calibration. Since these two wavelengths span the Argon-ion wavelengths of 514.5 nm and 488 nm, they were used along with this expression which is a

---

[48] W. Kaye and J.B. McDaniel, Applied Optics, 13, 1934 (1974).

[49] Cannell, et. al., Review of Scientific Instruments, 54, 973 (1983).

[50] Coumou, et. al., Journal of Colloid Science, 15, 408 (1960); Transactions of the Faraday Society, 60, 1726 (1960) and 60, 1539 (1964).

very mild function of wavelength:

$$R_U(90°) \cdot \frac{\lambda^4}{n^2} \cdot [(6 - 7 \cdot \rho_U)/(6 + 6 \cdot \rho_U)] \qquad \text{(SLS2-4)}$$

For benzene we used only Coumou's data to plot this expression at the classical wavelengths; for toluene we used Coumou's single value at 546 nm and Kaye's at 632.8 nm because Coumou's value at 436 nm seemed an outlier. Then the values of the expressions at 488 and 514.5 nm were interpolated, and R_U calculated. Finally, using this equation:

$$R_V(90°) = 2 \cdot R_U(90°)/(1 + \rho_U), \qquad \text{(SLS3-5)}$$

the Rayleigh ratio used in calibration, $R_V(90°) = R_C(90°)$, is determined. Here are the results:

488 nm	$R_V = R_C(90°)$	ρ_V	ρ_U
Toluene	40×10^{-6} cm^{-1}	0.31	0.47
Benzene	38.6×10^{-6}	0.27	0.43
514.5 nm	$R_V = R_C(90°)$	ρ_V	ρ_U
Toluene	32×10^{-6} cm^{-1}	0.31	0.47
Benzene	30×10^{-6}	0.26	0.41
532 nm	$R_V = R_C(90°)$	ρ_V	ρ_U
Toluene	29×10^{-6} cm^{-1}	0.31	0.47
Benzene	26×10^{-6}	0.26	0.41

Table SLS2-3: Absolute Rayleigh ratios at 488, 514.5, and 532 nm.

In recent years, the frequency doubled, solid state laser with a wavelength of 532 nm has come into use. Again, using a linear interpolation from the expressions at 632.8 nm and 488 nm, and similarly for interpolating ρ_U values, the values of $R_U(532)$ and $\rho_U(532)$ were determined. Using the equations above, $R_V(532)$ and $\rho_V(532)$ were calculated. All this is presented in Table SLS-3.

It is difficult to estimate accurate error bars for R_C at 488, 514.5, and 532 nm for toluene and benzene. But a conservative estimate is probably +/- 5%.

When doing SLS to determine M_W, it is a good idea to report the value of $R_C(90°)$ used for calibration. If two measurements disagree, it may be due to different choices of the Rayleigh ratio for the calibration liquid. Look for more recent R_C values in the literature, especially at 488, 514.5, and 532 nm. These values should be close to the ones given here but may be superior if better measurements of ρ_U or ρ_V are proven. If the same author has measured the Rayleigh ratio $R_C(90°)$ for toluene or benzene at 632.8 nm, they must be within +/- 2% of those above; otherwise, the author's measurements at other wavelengths are suspect. This last statement reflects the confidence in the HeNe measurements made by Kaye and Cannell ten years apart.

Appendix SLS3: Scattering Wave Vector, q

Referring to the scattering diagram, the magnitude of the incident wave vector $\mathbf{q_o}$ (bold) is designated q_o (not bold). The magnitude $q_o = 2\pi/\lambda = 2\pi n/\lambda_o$, where n is the refractive index of the medium in which the wave propagates, and λ_o is the wavelength of the beam in vacuum. Likewise, the magnitude of the scattered beam wave vector $\mathbf{q_s}$ (bold) is designated q_s (not bold). The magnitude $q_s = 2\pi/\lambda = 2\pi n/\lambda_s$, where, in general $\lambda_s \neq \lambda_o$.

Now consider the cases where neither the particle or molecule (including liquid molecules and everything dissolved in it) absorb at the incident wavelength. This corresponds to light scattering as the dominant interaction between the beam and the particle. In this case it turns out the difference in incident and scattered wavelength is less than one part in 10^8. Looked at from the point of view of frequency, the incident frequency is on the order of 10^{14} Hz and the shift in frequency (not proven here) is, for the translation of the smallest particles studied, 10^6. The frequency shift for larger particles is even smaller. Therefore $\lambda_s \approx \lambda_o$ to better than one part in 10^8. This defines quasi-elastic light scattering as used in SLS and DLS. In these experiments, no significant energy is absorbed by the system under study, no heating results.

The scattering wave vector $\mathbf{q}$ (don't confuse with the scattered wave vector, $\mathbf{q_s}$) is given by the difference: $\mathbf{q} = \mathbf{q_o} - \mathbf{q_s}$. Since the magnitudes q_s and q_o are equal (quasi-elastic), it follows that the triangle shown below is isosceles. Drop a vertical to the base, thus splitting the scattering angle into two equal parts ($\theta/2$), and forming two, equal right triangles.

The base of either right triangle x is given by its hypotenuse $2\pi n/\lambda_o$ times $\sin(\theta/2)$. Thus, the magnitude of the scattering wave vector is 2·x or:

$$q = \frac{4\pi n}{\lambda_o} \sin\frac{\theta}{2}$$

(SLS3-1)

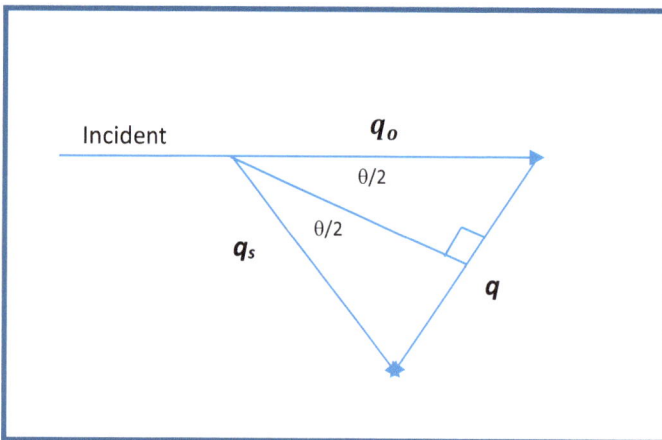

Scattered light, x-rays, neutrons, electrons—all scattering—shows the same result. The wavelength and refractive index (where they are defined) may change, but the equation remains the same.

Appendix SLS4: Photon Counting and Statistics

Photon Counting: Intensity

Intensity is the radiance in units of flux/area, where the flux is energy/time. Thus, radiance in MKS units is watts/m² and in CGS units is ergs/cm². When using a photomultiplier or photodiode detector in analog mode, the intensity is proportional to the photocurrent measured. If the photomultiplier tube is chosen properly, and if its output is modified with the correct amplifier-discriminator combination, individual photo pulses are registered per unit time and the count rate (counts/second or cps) corresponds to the intensity of scattered light. Likewise, avalanche photodiodes, APDs, when operated properly, produce a fluctuating stream of countable pulses.

Dark Current-Dark Count Rate

If there is a voltage applied to such detectors, even if no light falls on the active areas, a current can be detected. In the case of photon counting, it is called the dark count rate or DCR. The best detectors have low DCRs and are most useful in weak scattering situations. But if the DCR is constant, and measureable, it can be subtracted from the total measured intensity. The difference is the scattered intensity used to determine size and molecular weight and other nanoparticle properties.

For reference, a count rate > 500 kcps is considered high, 50–100 kcps is considered modest, and < 5 kcps is considered low. By contrast, DCRs are a few hundred cps or lower. These are approximate values and ranges.

Photon Counting Statistics: The Poisson distribution

The emission of a scattered photon as well as the single photon detector process is statistical. If the average count rate is 10^6 cps, then the average number of photons in 2 µs is 2. But the number registered in a 2 µs interval might be 0, 1, 2, 3, 4, 5, 6, 7…, because it follows Poisson statistics: countable, usually rare events per unit time. Photon, x-ray, and neutron counting, etc. follow Poisson statistics. When the average number of events per unit time is small, even zero is possible.

The Poisson distribution, which can be derived from the Binomial Distribution, follows this equation:

$$P_{\bar{n}}(n) = \frac{\bar{n}^n \cdot e^{-\bar{n}}}{n!} \tag{SLS4-1}$$

Where, the probability of n events per unit time given an average of $\bar{n}$ per unit time is $P_{\bar{n}}(n)$.

Figure SLS4-1 is a plot of the probabilities when $\bar{n}$ = 2, 5 and 8. The data points represent discrete values. As a visual guide, a smooth curve is drawn through the points, but they are discrete values only.

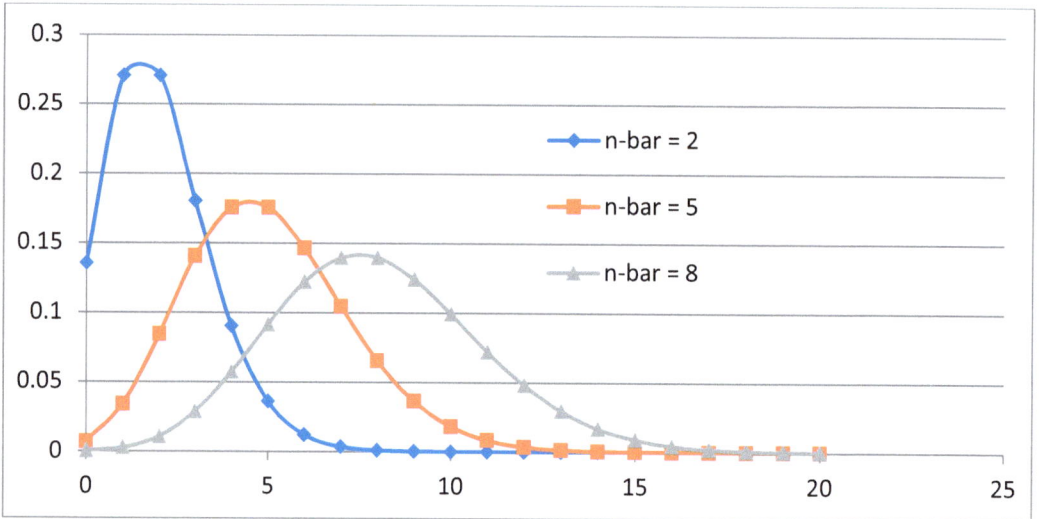

Figure SLS4-1: Poisson distribution plots.

Notice as $\bar{n}$ increases, the distribution becomes more symmetrical. In fact, it is shown in many textbooks on statistics that as $\bar{n}$ increases, the Poisson distribution approaches the Gaussian distribution. This is but one example of the Central Limit Theorem.

A Gaussian distribution is characterized by two values: the mean $\bar{n}$ and the standard deviation σ_G. These are independent of each other. A Poisson distribution is characterized only by one value, $\bar{n}$. Thus, it should not be surprising that its width, characterized by its standard deviation, is also a function of $\bar{n}$. In fact, for a Poisson distribution, $\sigma_P = \sqrt{\bar{n}}$. This leads directly to some simple consequences with useful experimental considerations.

If the dark count rate is 100 cps, then in any one second determination $\sigma_P = 10$ cps, and the relative error is 10%. You can reduce the error by counting for 10 seconds, accumulating approximately 1,000 total counts, resulting in $\sigma_P = 31.6$ cps and a 3.2% error. With good detectors, the dark count rate is very much lower than scattered light intensity and knowing it to this level of uncertainty is enough.

If you want to know the error in the measured scattered light intensity to 1%, you must count 10,000 total counts. That could be the result of 10 kcps for one second or 5 kcps for two seconds. The distribution and error apply to total counts not count rate.

For a count rate of 10^5 cps, normally easily attained for polymer solutions and particle suspensions, the random counting error is 0.32%.

Intensity Error: Corrected for Dark Count and Solvent Scattering

Remember, it is the *excess* scattered intensity that is related to mass, size and 2nd virial coefficient. Whether it is the Zimm, Debye, Berry, Fractal, Guinier or any of the other plots, it is either excess intensity or excess Rayleigh ratio that is plotted. And these are obtained by subtracting off solvent scattering. Solvent, of course, includes scattering from any dissolved small molecules and it too has a dark count embedded in it.

The excess can be written like this:

$$I_{ex} = I_{soln}^{meas} - I_{solv}^{meas} = (I_{soln} + I_{DC}) - (I_{solv} + I_{DC}) = I_{soln} - I_{solv} \qquad \text{(SLS4-2)}$$

In other words, when subtracting off solvent scattering, the dark count correction is automatically accounted for.

There is one measurement when it is not: When calibrating at 90° with, for example, a solvent like toluene whose Rayleigh ratio is known. In this case, there is no solvent to subtract, only the absolute scattering intensity of the toluene. Thus, subtracting the dark count is important.

Consider propagation of random error and what that means in this case when calibrating. The error is given by the square root of the sum of the squares of the errors in counting. The result looks like this:

$$I_{tol} = (I_{tol}^{meas} - I_{DC}) \pm \sqrt{\sigma_{I_{tol}^{meas}}^2 + \sigma_{I_{DC}}^2} \qquad \text{(SLS4-3)}$$

An example might be I_{tol}^{meas} = 30 kcps for one second or 30,000 total counts. Therefore $\sigma_{I_{tol}^{meas}}^2$ = 30,000. And picking a high dark count rate of 300 cps, assume, to reduce its error, it was counted for 10 seconds. The result is 3,000 counts with an error of 54.8 counts or 1.8%. Applied to 300 cps, that amounts to $\sigma_{I_{DC}}^2$ = 5.4². And the result is this:

$$I_{tol} = (30,000 - 300) \pm \sqrt{30,000 + 29} = 29,700 \pm 170 \ (0.6\%)$$

This is a small error and easily accounted for. A bigger problem arises when subtracting two numbers of closer magnitudes. Consider solution-solvent subtraction where the solvent scatters a lot. This difference will be smallest at the lowest concentration. An example is I_{soln} = 40,000 cps and I_{solv} = 30,000 cps. And assuming one second measurements:

$$I_{ex} = (40,000 - 30,000) \pm \sqrt{40,000 + 30,000} = 10,000 \pm 265 \ (2.65\%)$$

The individual errors in 40,000 and 30,000 are 200(0.5%) and 173(0.58%), but the error in the difference is much larger. There is a way around this, but it comes at a cost: much longer measurement time and increased likelihood of dust moving through the beam. Increase the duration from one second to 10. The error of 2.65% is reduced to 0.84%.

Intensity Error: Dead-Time Correction

Detectors cannot respond to photo pulses instantaneously. After the arrival and registration of one pulse, there is a certain time after which the detector can't respond. If during this time another pulse arrives, it is not registered. This time is called the dead-time, τ_d. With good PMTs and APDs it might be 25 ns; with poorer response times, it might be 125 ns.

For low count rates, this is not significant; whereas, for high count rates it can be. A correction should be applied to the measured value, I_M, to produce the true value, I_T. A relatively simple expression, good perhaps up to a 15% correction is this:

$$I_T = \frac{I_M}{1 - \tau_d \cdot I_M}$$

(SLS4-4)

The following table illustrates the consequences for two different dead times.

I_M (cps)	$I_T(\tau_d = 25\ ns)$	$I_T(\tau_d = 125\ ns)$
10^2 dark count rate	100	100
10^3 weak count rate	1,000	1,000
10^4 acceptable	10,003 (0.03%)	10,013 (0.13%)
10^5 strong count rate	100,251 (0.25%)	101,266 (1.27%)
10^6 strong count rate	1,025,641 (2.6%)	1,142,857 (14.3%)

Table SLS3-1: Effect of dead-time on true count rate vs measured count rate.

If you want intensities correct to better than 1%, with a dead-time correction of 25 ns, you don't need to worry until the count rate is well past a few hundred thousand counts/second. But if you have a slower detector, with 125 ns dead-time, you need to make the correction starting around 100 kcps. The maximum measured count rate beyond which the correction should not be applied is around 1 Mcps for a detector with $\tau_d = 125$ ns. In this case, arrange laser power, apertures, or concentrations to achieve a lower count rate where the correction does apply.

Summary

Photon counting in light scattering removes the errors associated with analog measurements: calibration, drift, etc. Select the best single photon response detector with the lowest dark count rate and dead-time. Make such corrections. When using such a detector in DLS, know what the after-pulsing limitations are for the specific detector. Either stay above the delay time limit, above which after pulsing is insignificant, or consider cross correlation.

Appendix SLS5: Depolarization Corrections

Corrections for Depolarization

For optically isotropic molecules, the polarization of the scattered light is parallel to the polarization of the incident light. The most common configuration with a laser source is incident light that is linearly polarized perpendicular to the scattering plane. This is commonly called vertical polarization. For many, large, flexible-chain polymers depolarization effects are negligible, and, as a result, no polarization corrections are necessary. M_w, R_g, and A_2 are correct as calculated from equations II-13, II-14 and II-18. Many solvents are anisotropic and depolarize the scattered light significantly; however, these effects are negated upon subtracting the solvent scattering from that of the solution.]

For optically anisotropic molecules, the polarization of the scattered light has components both parallel and perpendicular to the incident polarization. For short, flexible-chain polymers and chains in or near the rigid rod limit, depolarization effects are not negligible, and, as a result, polarization corrections are necessary. Under these circumstances M_w, R_g, and A_2, as used in the equations in Chapter II, should be considered apparent (denoted with "app" subscript), not true values.

To make the corrections, two limiting cases of the worm-like chain model must be distinguished: flexible chain and rigid rod. In addition, each of these two cases can be measured in either of two common experimental setups: measurements with linearly polarized incident light perpendicular to the scattering plane and no analyzer I_{uv}; or, measurements with an analyzer set to pass only scattered light with polarization parallel to the incident polarization I_{vv}.

For I_{uv} measurements on flexible chain molecules:

$$M_{w,app} = M_w \cdot [1 + (7/5)\delta^2] = M_w \cdot [(3 + 3\rho_v)/(3 - 4\rho_v)] \tag{SLS5-1}$$

$$A_{2,app} = A_2/[1 + (7/5)\delta^2] = A_2/[(3 + 3\rho_v)/(3 - 4\rho_v)] \tag{SLS5-2}$$

$$R_{g,app} = R_g/[1 + (7/5)\delta^2]^{0.5} = R_g/[(3 + 3\rho_v)/(3 - 4\rho_v)]^{0.5} \tag{SLS5-3}$$

For I_{uv} measurements on rigid rod molecules:

$$R_{g,app} = R_g \cdot \{[1 - (4/5)\delta + (47/35)\delta^2]/[1 + (7/5)\delta^2]\}^{0.5} \tag{SLS5-4}$$

For I_{vv} measurements on flexible chain molecules:

$$M_{w,app} = M_w \cdot [1 + (4/5)\delta^2] = M_w \cdot [3/(3 - 4\rho_v)] \tag{SLS5-5}$$

$$A_{2,app} = A_2/[1 + (4/5)\delta^2] = A_2/[3/(3 - 4\rho_v)] \qquad \text{(SLS5-6)}$$

$$R_{g,app} = R_g/[1 + (4/5)\delta^2]^{0.5} = R_g/[3/(3 - 4\rho_v)]^{0.5} \qquad \text{(SLS5-7)}$$

For I_{vv} measurements on rigid rod molecules:

$$R_{g,app} = R_g \cdot \{[1 - (4/5)\delta + (4/7)\delta^2]/[1 + (4/5)\delta^2]\}^{0.5} \qquad \text{(SLS5-8)}$$

In these equations δ is the molecular anisotropy and ρ_v is the depolarization ratio. The depolarization ratio is an experimentally measured parameter. It is operationally defined as:

$$\rho_v = \Delta I_{hv}(90°)/\Delta I_{vv}(90°) \qquad \text{(SLS5-9)}$$

Where, $\Delta I_{hv}(90°)$ is the <u>excess</u> scattered light intensity measured at $\theta = 90°$ with polarization perpendicular to the incident polarization, and $\Delta I_{vv}(90°)$ is the <u>excess</u> scattered light intensity measured at $\theta = 90°$ with polarization parallel to the incident polarization. The excess is defined by the difference between the value measured with the solution and the value measured with the solvent. Since many common solvents can depolarize the light significantly, it is extremely important to use the excess values.

The molecular anisotropy can be written in terms of the experimentally determined depolarization ratio as:

$$\delta^2 = 5\rho_v/(3 - 4\rho_v) \qquad \text{(SLS5-10)}$$

A polarization analyzer on the detector is best for determination of ρv. Alternatively, literature values of ρ_v can be input, and its effect on recalculated values M_w, R_g, and A_2 determined.

For most polymer samples the depolarization corrections are negligible. This is so because the normally random orientation of dipole moments from the anisotropic monomeric units in a polymer cancels as the molecular weight increases. In fact, it is often stated that if the measured depolarization ratio is not negligible, the measurement is probably wrong. The older literature values of depolarization ratios are suspect. Be careful in applying these corrections. If in doubt, do not apply them. With ρ_v equal to zero, there should be no effect on the calculation of M_w, R_g, or A_2 even if the flexible chain or rigid rod choices are selected.

For rod-like polymer molecules, for short-chain molecules where end-groups may contribute significantly to depolarization, ρ_v may not be negligible, and corrections are necessary.

Appendix DLS1: Scattering Volume, Concentrations & Particle Interactions

Scattering Volume & Minimum Concentration

In a typical DLS experiment, one using lenses and pinholes, a nominal 1 mm laser beam is focused to approximately 100 μm. The focusing is such that, over the length of the detected region, the beam can be modeled as a cylinder. [Focused laser beams are hyperbolas of revolution, but one usually chooses the focusing optics such that the cylindrical approximation is reasonable.] A common detection scheme might look at a 200 μm length of the cylinder at a 90° scattering angle. The resultant scattering volume $V_{sca}(90°) \approx 1.6 \times 10^{-6}$ cm³. At other angles, with the optics fixed, the length of the scattering volume observed is larger by an amount that varies inversely with the sin of the scattering angle. Thus $V_{sca}(\theta) \approx 1.6 \times 10^{-6}$ cm³/sin(θ), doubling in volume at 30° and 150°, quadrupling in volume at 15°.

In statistics, a common assumption is that at least 30 values are needed for Gaussian statistics to apply. And in DLS theory, for the Siegert relationship to apply, Eq. (III-2), the phase of the scattered light must be a Gaussian random variable. Thus, at least 30 particles in the scattering volume must contribute to the total scattered light. Assume that the minimum number of scattering particles per scattering volume is 30, though experimentally one finds a minimum more like 10 times that number is required to build up the ACF in a reasonable time. What does this mean for the concentration?

Let $C_n \equiv$ (Number of Particles/Unit Volume). At 90°, $C_n(min) \approx 30/1.6 \times 10^{-6}$ cm³ $\approx 1.9 \times 10^{+7}$ particles/cm³. This means that DLS measurements are ensemble average measurements over large numbers of particles; they are not single particle measurements. For this reason alone, a DLS determination of the particle size distribution is no substitute for a single particle counter, if an accurate measurement of the number distribution is required.

More often the concentration of interest is either the volume fraction $\phi_v \equiv$ (Volume of Particles/Unit Volume), which, for a monodisperse dispersion is given by $\phi_v \equiv v_p \cdot C_n$, or the weight/volume concentration $C_w \equiv$ (Mass of Particles/Unit Volume) $= \varrho_p \cdot \phi_v$. Here $v_p = (\pi/6) \cdot d^3$, the volume of a single particle, and ϱ_p is the particle density of a spherical particle of diameter d. Minimum concentrations for a range of sizes is then given by:

d (nm)	ϕ_v(min)	C_w(min, $\varrho_p = 1$ g/cm³)
10	10^{-11} v/v = 10^{-9} %v/v	10^{-8} w/v* = 10^{-6} %w/v
100	10^{-8} v/v = 10^{-6} %v/v	10^{-5} w/v = 10^{-4} %w/v
1000	10^{-5} v/v = 10^{-3} %v/v	10^{-3} w/v = 10^{-1} %w/v

* w/v refers to mg/mL

Table DLS1-1: Estimates for minimum v/v and w/v concentrations vs. size for DLS.

Experience with real samples suggests a minimum volume concentration of 10^{-5} %v/v for 100 nm particles in accordance with the suggestion above that while 30 particles may be the minimum for Gaussian statistics and the Siegert relationship to apply, something like 10 times that number is required for realistic measurements.

For concentrations lower than those in the table, one can expect number fluctuations. See the very brief discussion on this subject in the section on data analysis and consult the references at the beginning of that section.

Scattering Volume & Maximum Concentration

Another aspect of DLS applied to particle sizing may be illustrated using the table above. To calculate particle size from the diffusion coefficient, it is assumed that the Stokes-Einstein law applies. Therefore, it is assumed that the particles do not interact. This is certainly true in dilute suspensions. In the case of particles electrostatically stabilized, interactions may be minimized, even at higher particle concentrations, by shielding the particles using a diluent consisting of 10 to 30 mM, 1:1 electrolyte. Note, however, adding too much salt can have the opposite effect: collapsing of the electrical double layer, resulting in flocculation or aggregation. But this effect normally occurs well above 100 mM for 1:1 electrolyte. See the zeta potential appendices.

As a crude guess, let us assume that when the center-to-center distance between particles, L, is greater than 20 times the particle diameter, d, that they do no interact. A crude model of a suspension might assume, on average, something less than the 12 nearest neighbors in a close-packed crystalline structure such as a face-centered cubic or a hexagonal closest packed unit cell. Such a model leads to a simple relationship between the volume fraction and the ratio of d/L. Depending on the number of nearest neighbors, the relationship is given by $\phi_v(max) = b \cdot (d/L)^3$, where b = 0.74 for a closest packed crystalline structure with 12 nearest neighbors, and b = 0.5 with a loose, amorphous structure with 6 nearest neighbors.

Assuming d/L < 0.05 is reasonable for noninteracting particles, this leads to a $\phi_v(max)$ of 6.3×10^{-3} %v/v. This value is only 6.3 times as large as the $\phi_v(min)$ for a 1 µm particle, assuming a minimum of 30 particles in the scattering volume. So, the range of concentrations from the minimum to the maximum is 6.3. However, if the minimum number of particles is more like 300, as it is for much smaller particles, then it is literally impossible to satisfy the two opposing criteria.

Thus, while it is easy to satisfy the opposing concentration criteria of at least 30 to 300 particles in the scattering volume but not so many that the particles interact, for the smaller diameter particles, it becomes increasingly difficult if not impossible to do so as particles above 1 µm or so are considered.

Particles above this size range scatter more and more in the forward direction, making it difficult to measure since alignment and dust are significant problems in the forward direction.

Also, larger particles may sediment during the measurement. They also diffuse more slowly, requiring much longer measurement times. Yet, the violation of the Siegert relationship is as fundamental to the breakdown of DLS above this size range as any of the other effects. For these reasons, particle sizing above, very approximately, 1 μm using DLS is not easy to justify.

[Note: While one could increase the focused beam diameter to allow a greater number of particles in the scattering volume, the power per coherence area decreases, and this leads to much reduced intercept/baseline ratios. Repeatability suffers and it becomes nearly impossible to obtain reliable second moments as the rise above the baseline in the tail of the ACF is barely perceptible above the noise. In addition, increasing the beam diameter increases the chances of a dust particle crossing the beam during the measurement. And when it does, it has a longer illuminated path in which to travel.]

Appendix DLS2: The Z-average Diameter vs. 1st Cumulant Diameter

$\overline{M_n}$, $\overline{M_w}$ and $\overline{M_z}$ are long established terms in static light scattering from polymer solutions. They correspond to averaging M_i, the molecular weight of a thin slice of the molecular weight distribution, by n_i, M_i or I_i. These represent weighting by the number of molecules per unit volume, by the mass or the scattered intensity of the ith class, respectively. The intensity is angular dependent, but the z-average is not. Thus, the z-average occurs when one refers to extrapolation to zero angle or to very small particles ($R_g/\lambda<<1$), both cases where the angular dependent part, $P_i(\theta) \rightarrow 1$. The radius of gyration, more properly the root-mean-square radius, is also a z-average $\overline{(R_g)}_z$ and it too is obtained using the intensity as the weighting factor.

It was shown in the text that the concentration of the ith class is given by $c_i = n_i \cdot M_i$, and that the scattered intensity I_i is proportional to $c_i \cdot M_i \cdot P_i(\theta) = n_i \cdot M_i^2 \cdot P_i(\theta)$. The intensity averaging of any quantity x, is therefore given by:

$$\overline{X_I} = \Sigma n_i \cdot M_i^2 \cdot P_i(\theta) \cdot x_i) \,/\, \Sigma n_i \cdot M_i^2 \cdot P_i(\theta) \qquad \text{(DLS2-1)}$$

And, for spheres where $M_i^2 \propto d_i^6$:

$$\overline{X_I} = \Sigma n_i \cdot d_i^6 \cdot P_i(\theta) \cdot x_i) \,/\, \Sigma n_i \cdot d_i^6 \cdot P_i(\theta) \qquad \text{(DLS2-2)}$$

By letting $x_i = d_i$, and for the cases where $P_i(\theta) = 1$, the z-average for a particle size distribution is given by:

$$\overline{d_z} = \Sigma n_i d_i^7 \,/\, \Sigma n_i d_i^6 \qquad \text{(DLS2-3)}$$

In DLS, the 1st cumulant diameter arises from the intensity weighting of the assumed translational diffusion coefficient, D, not the diameter d. Given the inverse relationship between D and d, these two types of averaging can never be equal even if $P_i(\theta) = 1$.

From the 1st cumulant one gets:

$$\overline{D_I} = \Sigma I_i \cdot D_i / \Sigma I_i = \Sigma n_i \cdot M_i^2 \cdot D_i \cdot P_i(\theta) \,/\, \Sigma n_i \cdot M_i^2 \cdot P_i(\theta) \qquad \text{(DLS2-4)}$$

And, for spheres where $M_i^2 \propto d_i^6$ and $D_i \propto 1/d_i$:

$$\overline{(1/d)_I} = \Sigma n_i \cdot d_i^6 \cdot P_i(\theta) \cdot (1/d_i) \,/\, \Sigma n_i \cdot d_i^6 \cdot P_i(\theta) \qquad \text{(DLS2-5)}$$

The true z-average obtains when $P_i(\theta) = 1$ and:

$$\overline{(1/d)_z} = \Sigma n_i \cdot d_i^5 \ / \ \Sigma n_i \cdot d_i^6 \qquad\qquad \text{(DLS2-6)}$$

Inverting this last expression gives the DLS diameter from the 1st cumulant as

$$\overline{d_{1st\ cumulant}} = \Sigma n_i d_i^6 / \Sigma n_i d_i^5, \ \text{ when } P_i(\theta) = 1. \qquad\qquad \text{(DLS2-7)}$$

This is also called the *apparent or effect diameter* from DLS.

And for diameters large enough to show angular dependence, the first cumulant diameter is also angular dependent and given by:

$$\overline{d_{1s\ cumulant}}(\theta) = \Sigma n_i d_i^6 \, P_i(\theta) / \Sigma n_i d_i^5 \, P_i(\theta) \ , \ P_i(\theta) \neq 1 \qquad\qquad \text{(DLS2-8)}$$

This last expression tells us why measurements at high scattering angles are not preferred over small angles. For non-Rayleigh particles, large enough so $P_i(\theta) \neq 1$, the first cumulant is a function of angle.

Summary: The two expressions in the shaded boxes are never equal, independent of the argument about $P_i(\theta)$. Therefore, it is never true that the z-average diameter equals the diameter calculated from the 1st cumulant using DLS. Sadly, the myth persists.

Appendix DLS3: Limitations of DLS in Particle Sizing Using Bi-modals and Cumulants

A few interesting consequences of d^6 weighting on scattered intensity and of the limits of DLS for determining particle size distribution can be demonstrated using a bimodal made up of only two peaks, each peak consisting of only one size.

From Eqs. (III-8) and (III-10), the first cumulant in Γ-space can be shown to be:

$$\overline{\Gamma} = \frac{I_1 \Gamma_1}{I_1 + I_2} + \frac{I_2 \Gamma_2}{I_1 + I_2} \tag{DLS3-1}$$

where

$$\frac{I_1}{I_2} = \left(\frac{N_1}{N_2}\right) \cdot \left(\frac{d_1}{d_2}\right)^6 \cdot \left(\frac{P_1}{P_2}\right) \tag{DLS3-2}$$

with N the concentration as the number of particles per unit volume, d the diameter, and P the angular scattering factor for each of the two sizes. See Chapter II for details.

Using Eqs. (III-3), (III-4) and (III-5) for translational diffusion of dilute spheres, it is simple to show:

$$\frac{\overline{\Gamma}}{\Gamma_1} = \frac{d_1}{d_{DLS}} \tag{DLS3-3}$$

From Eq. (III-11) it is easy to show that the second cumulant μ_2 is given by:

$$\mu_2 = \overline{\Gamma^2} - (\overline{\Gamma})^2 \tag{DLS3-4}$$

This should be familiar to anyone who has studied basic statistics: the variance is equal to the average of the square minus the square of the average.

It also follows from Eqs. (III-8) and (III-10) that the average of the square of Γ is given by:

$$\overline{\Gamma^2} = \frac{I_1 \Gamma_1^2}{I_1 + I_2} + \frac{I_2 \Gamma_2^2}{I_1 + I_2} \tag{DLS3-5}$$

Subtracting the square of Eq. (DLS3-1) from Eq. (DLS3-5) and simplifying yields:

$$\mu_2 = \frac{I_1 \cdot I_2}{(I_1 + I_2)^2} \cdot (\Gamma_1 - \Gamma_2)^2 \tag{DLS3-6}$$

If the two sizes are equal, the two Γ's are equal and μ_2 is zero. The distribution is mono-modal.

Now the polydispersity index (PDI), also called Poly is the variance divided by the mean squared. Again, from statistics, this is the relative variance and therefore has no units. It is a relative measure of distribution width with small values indicating narrow distributions and larger values broader distributions. Experience with standards such as polystyrene show that Poly ≤ 0.025 is consistent with very narrow distributions. The d_{DLS} obtained agrees within experimental error with labelled values that are often from image analysis. When two weighted averages are equal, the distribution is monodisperse. When they are equal within experimental error, the distribution is either monodisperse or very narrow.

Combining Eqs. (DLS3-5) and (DLS3-6) yields the following for Poly:

$$\text{Poly} = \frac{\mu_2}{\overline{\Gamma}^2} = \frac{I_2}{I_1} \cdot \frac{\left(1-\frac{d_1}{d_2}\right)^2}{\left(1+\frac{I_2}{I_1}\cdot\frac{d_1}{d_2}\right)^2} \tag{III-7}$$

Examples and Consequences

We can now calculate d_{DLS}, the diameter calculated from the first cumulant, and Poly, a measure of relative width (variance) from the 1st and 2nd cumulants. Combining Eqs. (III-1), (III-2) and (III-3) results in:

$$d_{DLS} = d_1 \cdot \frac{1+\left(\frac{N_2}{N_1}\right)\cdot\left(\frac{d_2}{d_1}\right)^6\cdot\left(\frac{P_2}{P_1}\right)}{1+\left(\frac{N_2}{N_1}\right)\cdot\left(\frac{d_2}{d_1}\right)^5\cdot\left(\frac{P_2}{P_1}\right)} \tag{III-8}$$

And combining Eqs. (III-2) with (III-7) results in:

$$\text{Poly} = \left(\frac{N_2}{N_1}\right) \cdot \left(\frac{d_2}{d_1}\right)^6 \cdot \left(\frac{P_2}{P_1}\right) \cdot \frac{\left(1-\frac{d_1}{d_2}\right)^2}{\left(1+\left(\frac{N_2}{N_1}\right)\cdot\left(\frac{d_2}{d_1}\right)^5\cdot\left(\frac{P_2}{P_1}\right)\right)^2} \tag{III-9}$$

Using N_2 as the independent variable and therefore $N_1 = 100 - N_2$, we are consistent with the first paragraph of this appendix wherein it is assumed there are only two sizes, each representing a peak or mode in the distribution. Furthermore, the N's now represent the percent by number/volume in their respective peaks. As such, they will vary from 0 to 100. But remember they are not absolute numbers but percentages. For example, with $N_1 = 1$, the actual number of particles depends on the total. If it were 10^6, then $N_1 = 10,000$. Thus, unlike "dust", which is defined as a few large particles, the concentrations here represent populations with enough particles to contribute to the measurement.

Example A: Two peaks only 25% apart (100 nm/80 nm)

DLS is not a high-resolution particle sizing technique. The ACF is a result of a sum (or integral) over exponentials. As such, it is not easily deconvoluted. For peaks closer than about 2:1, often a single, peak results. Table DLS3-1 shows a calculated example covering a range from 1% to 99% by number.

Notice that the polydispersity values are all ≤ 0.012. Thus, the normal interpretation is that this is a monodisperse distribution at the calculated d_{DLS} value. It is the wrong interpretation. The correct one is that this distribution is as narrow as DLS can measure. Individual peaks could be measured using other techniques: a disc centrifuge, a field-flow fraction device, a column hydrodynamic fractionator, or single particle counter. Yet, each of these has its own drawbacks, but are all considered high resolution.

Polystyrene Latex in Water
Temperature 25 °C
Scattering Angle 90°
Wavelength 632.8 nm (HeNe laser)
Particle sizes: $d_1 = 80$ nm, $d_2 = 100$ nm
Angular scattering factors from Mie theory: $P_1 = 0.8907$, $P_2 = 0.8345$

% Num N_2	% Num N_1	d_{DLS} in nm	Poly	% Conc c_2	% Conc c_1	% Int I_2	% Int I_1
1	99	80.6	0.001	1.9	98.1	3.5	96.5
2	98	81.1	0.003	3.8	96.2	6.8	93.2
5	95	82.6	0.006	9.3	90.7	15.8	84.2
10	90	84.8	0.009	17.8	82.2	28.4	71.6
20	80	88.3	0.012	32.8	67.2	47.2	52.8
50	50	94.8	0.010	66.1	33.9	78.1	21.9
80	20	98.4	0.004	88.7	11.3	93.5	6.5
90	10	99.3	0.002	94.6	5.4	97.0	3.0
95	5	99.6	0.001	97.4	2.6	98.5	1.5
98	2	99.9	0.000	99.0	1.0	99.4	0.6
99	1	99.9	0.000	99.5	0.5	99.7	0.3

Table DLS3-1: Effect of concentration on cumulant results for 80 and 100 nm PS/Water.

Notice also the effect of d^6 on the % by mass concentrations (C's) and intensities (I's). At 50:50 by number, mass concentrations are almost 2:1 and the intensity concentrations are nearly 4:1 in favor of the larger particles. Because of d^6, larger particles contribute more to scattering than smaller particles, and therefore more to the ACF, except in the rare cases where the angular scattering coefficients (Mie values, shown as P's here) are close to zero. In those rare cases, it is possible to see reversals in the normal trends.

Example B: Two peaks 2:1 apart (20 nm/10 nm)

Gold nanoparticles are used in a variety of situations and are commercially available. But Au nanoparticles are hard to keep apart. One manufacturer sells a 10-nm product with the specification that no more than 8% are doublets. In one example, d_{DLS} of about 16 nm was found under a variety of conditions. How is this to be explained? See Table DLS3-2.

If just 4% to 5% by number (N2) are doublets (20 nm), it is enough to explain the result. Such aggregation is consistent with the manufacturer's specification that no more than 8% are doublets. In fact, using DLS instead of electron microscopy, one can much more quickly estimate the percent of doublets (assuming no other multiplets) using the measured d_{DLS}.

The Poly values generally signify this is not a unimodal distribution. And the effects of d^6 are quite evident. With just a 1% by number (N2) of the larger particle, d_{DLS} is shifted 24% higher than 10 nm. That 1% by number represents 61% by scattered intensity. If the doublet concentration by number is only 1 in 1,000 (N2 = 0.1), then d_{DLS} is close to the expected 10 nm.

Polystyrene Latex in Water
Temperature 25 °C
Scattering Angle 90°
Wavelength 632.8 nm (HeNe laser)
Particle sizes: $d_1 = 10$ nm, $d_2 = 20$ nm
Angular scattering factors from Mie theory: $P_1 = 0.9993$, $P_2 = 0.9974$

% Num N_2	% Num N_1	d_{DLS} in nm	Poly	% Conc c_2	% Conc c_1	% Int I_2	% Int I_1
8	92	17.4	0.097	41.0	59.0	84.7	15.3
7	93	17.1	0.104	37.6	62.4	82.8	17.2
6	94	16.7	0.110	33.8	66.2	80.3	19.7
5	95	16.3	0.117	29.6	70.4	77.1	22.9
4	96	15.7	0.122	25.0	75.0	72.7	27.3
3	97	15.0	0.125	19.8	80.2	66.4	33.6
2	98	13.9	0.119	14.0	86.0	56.6	43.4
1	99	12.4	0.092	7.5	92.5	39.2	60.8
0.5	99.5	11.4	0.060	3.9	96.1	24.3	75.7
0.1	99.9	10.3	0.015	0.8	99.2	6.0	94.0

Table DLS3-2: Effect of concentration on cumulant results for 10 nm and 20 nm Au/Water.

This is an example of using DLS to check on monodispersity in particle reference standards, globular proteins, monoclonal antibodies, and any other particles where monodispersity is either desired or should be obtained.

Example C: Two peaks 5:1 apart (250 nm/50 nm)

Polystyrene Latex in Water

Temperature 25 °C

Scattering Angle 90°

Wavelength 632.8 nm (HeNe laser)

Particle sizes: $d_1 = 50$ nm, $d_2 = 250$ nm

Angular scattering factors from Mie theory: $P_1 = 0.955$, $P_2 = 0.2695$

% Num N_2	% Num N_1	d_{DLS} in nm	Poly	% Conc c_2	% Conc c_1	% Int I_2	% Int I_1
1	99	229.8	0.290	55.8	44.2	97.8	2.2
2	98	239.5	0.160	71.8	28.2	98.9	1.1
5	95	245.8	0.066	86.8	13.2	99.6	0.4
10	90	248.0	0.032	93.3	6.7	99.8	0.2
20	80	249.1	0.014	96.9	3.1	99.9	0.1
50	50	249.8	0.004	99.2	0.8	100.0	0.0
80	20	249.9	0.001	99.8	0.2	100.0	0.0
90	10	250.0	0.000	99.9	0.1	100.0	0.0
95	5	250.0	0.000	100.0	0.0	100.0	0.0
98	2	250.0	0.000	100.0	0.0	100.0	0.0
99	1	250.0	0.000	100.0	0.0	100.0	0.0

Table DLS3-3: Effect of concentration on cumulant results for 50 nm and 250 nm PS/Water.

With even 95% by number of the 50 nm particles (N1), the results on d_{DLS} are within 1.7% (250/245.8) of 250 nm. It is as if in a sea of small particles, much larger particles dominate. This is also evident by the fact that the intensity of the larger particles (I2) is nearly 100% of the total scattered intensity. Only the Poly of 0.066 hints at a broader distribution.

Summary: Due mainly to the d^6 weighting of the intensity, small particles contribute very little to the ACF and the calculated results unless they are in overwhelming number or if the sizes are somewhat close. While this limits DLS for distinguishing small particles in the presence of large ones, it also makes it easy to determine large ones in the presence of small ones. DLS is an excellent monitor of early onset aggregation.

Appendix Z1:   Ionic Strength

The ionic strength, on a concentration basis, is defined as follows:

$$I_c = \tfrac{1}{2} \sum c_i \cdot z_i^2 \qquad\qquad (Z1\text{-}1)$$

Where z_i is the valence (typically, 1, 2, 3, or 4) of the ith ion and c_i is the concentration of the ith ion in solution.

[Note: This definition does not include the charged particles themselves. For suspensions, it is strictly true that the charged particle is not in solution (they are in suspension); however, for charged polymers of colloidal size, and for proteins, such particles are in solution. Furthermore, whether in solution or not, all charged particles contribute to the overall conductivity. Typical measurements, however, are carried out at sufficiently high, added salt concentration (10^{-3} to 10^{-2} Molar or more) so that ignoring any contribution of charged, colloidal-sized particles in the calculation of I_c does not affect significantly the results, and the measured conductivity is determined, within experimental error, only by the added salt concentration including acids and bases.]

The units of I_c are the units of c_i, mol/m³ in the SI system of units and molar (mol/L or M) in the older chemical literature system of units. Since 1 L is now defined as exactly 10^{-3} m³ or 1 dm³, a 1 M salt solution is equal to a 10^3 mol/m³ or a 1 mol/dm³ solution.

Typical added salts include NaCI (1:1), $CaCl_2$ (2:1), and $Al_2(SO_4)_3$ (3:2). (In addition, samples prepared from so-called city water include smaller amounts of +2 and +3 ions such as Ba^{+2} and Al^{+3}) Electrolytes are classified according to the charge on the ions. For example, KCl is a 1:1 electrolyte, $CuSO_4$ is a 2:2 electrolyte, and Na_2SO_4 is a 1:2 electrolyte. When these salts are added in the concentration range of 10^{-2} M or less, they completely dissociate into their separate ions.

The calculation of ionic strength then proceeds as follows:

Electrolyte	Calculation	Ionic Strength
1:1	$I = \tfrac{1}{2} [c \cdot 1^2 + c \cdot 1^2]$	c
1:2 or 2:1	$I = \tfrac{1}{2} [2 \cdot c \cdot 1^2 + c \cdot 2^2]$	$3 \cdot c$
2:2	$I = \tfrac{1}{2} [c \cdot 2^2 + c \cdot 2^2]$	$4 \cdot c$
1:3 or 3:1	$I = \tfrac{1}{2} [3 \cdot c \cdot 1^2 + c \cdot 3^2]$	$6 \cdot c$
3:3	$I = \tfrac{1}{2} [c \cdot 3^2 + c \cdot 3^2]$	$9 \cdot c$
2:3 or 3:2	$I = \tfrac{1}{2} [3 \cdot c \cdot 2^2 + 2 \cdot c \cdot 3^2]$	$15 \cdot c$

Table Z1-1: Ionic strength in terms of salt concentration for a variety of electrolytes.

Since the double layer thickness varies inversely as the square root of the ionic strength, even relatively small amounts of multivalent ions can decrease the thickness by factors of 2 and 3 readily. This leads to a drastic decrease in the distance over which the electrostatic repulsive force is significant. And, as a result, increases the chance that the particles will aggregate.

It is much easier to work in low to modest salt concentrations, say 1 mM to 20 mM, because not much heating occurs due to the passage of current. Physiological saline, 0.9% wt/vol NaCl is 155 mM, modestly high salt. Ocean water, brines and electroplating fluids are typically very high in salt and so in ionic strength. Here is a table of ions and concentrations for a model sea water solution used in research.

Ions	Conc. (ppm)	Conc. (Molar)
Na^{+1}	18,300	0.796
Cl^{-1}	32,200	0.908
HCO_3^{-1}	120	0.00197
Ca^{+2}	650	0.0162
Mg^{+2}	2,110	0.0868
SO_4^{-2}	4,290	0.0447

Table Z1-2: Common model sea water salt solution.

The ionic strength is given by:

$$I = \frac{1}{2}[(0.796 + 0.908 + 0.00197) \cdot 1^2 + (0.0162 + 0.0868 + 0.0447) \cdot 2^2]$$
$$= 1.15 \; Molar$$

That is 1,150 mM or 7.4 times saltier than saline. At such a high concentration, the double layer thickness is 0.28 nm. It is so small that the standard understanding of zeta potential has lost its meaning. Whereas the operationally defined electrophoretic mobility can still be measured, the subsequently calculated zeta potential may not be of value.

Appendix Z2:   Double Layer Thickness

Definition

The double layer thickness is defined by the following equation:

$$\kappa^{-1} = \{(\varepsilon \cdot k_B \cdot T)/(2e^2 \cdot 1000 \cdot I_c \cdot N_{avo})\}^{0.5} \qquad (Z2\text{-}1)$$

Where,

ε	=	Permittivity of Liquid
k_B	=	Boltzmann's Constant, $1.3807 \times 10^{-23}\,\text{J}\cdot\text{K}^{-1}$
T	=	Temperature in Kelvin (K)
e	=	Electronic Charge in Coulombs, 1.6022×10^{-19} C
I_c	=	Ionic Strength in units of mol/dm^3 (mol/L or M, molar)
N_{avo}	=	Avogadro's Number, $6.0221 \times 10^{23}\,\text{mol}^{-1}$

The permittivity of the liquid is equal to the product of $\varepsilon_r \cdot \varepsilon_0$, where ε_r, the relative permittivity, also called the dielectric constant, is 78.54 for water at 298.15 K. The permittivity of vacuum, ε_0, is 8.8542×10^{-12} F·m^{-1} where 1 Farad= $1 \cdot C^2 \cdot J^{-1}$. The extra factor of 1000 in the denominator converts I_c from mol/dm^3 to mol/m^3. Substituting these values into the equation, one obtains:

$$\kappa^{-1} = \{9.2591 \times 10^{-20}/I_c\}^{0.5} \text{ in meters, or} \quad \kappa^{-1} = 0.3043/I_c^{0.5} \text{ in nanometers} \qquad (Z2\text{-}2)$$

The double layer thickness decreases inversely as the square root of the ionic strength. The ionic strength depends on the concentration of free ions and on their charge. See Appendix Z1. For simple salts, the double layer thickness in water at 25 °C as a function of concentration is given in the following table:

Double Layer Thickness in Nanometers

Conc. (M) →	10^{-7}	10^{-6}	10^{-5}	10^{-4}	10^{-3}	10^{-2}	10^{-1}
Electrolyte	Double Layer Thickness in nm						
1:1	962	304	96.2	30.4	9.62	3.04	0.962
1:2, 2:1	555	176	55.5	17.6	5.55	1.76	0.555
2:2	481	152	48.1	15.2	4.81	1.52	0.481
1:3, 3:1	393	124	39.3	12.4	3.93	1.24	0.393
3:3	321	101	32.1	10.1	3.21	1.01	0.321
2:3, 3:2	248	78.5	24.8	7.85	2.48	0.785	0.248

Interpretation

Much can be learned from such a table. First, in deionized water, here the only free ions are H^+ and OH^-, the concentration is 10^{-7} M, and the double layer thickness is 962 nm, nearly 1 μ. This is the distance over which the repulsive forces are significant. Second, add a little KCl, a 1:1 salt, and, at a concentration of 10 mM (millimolar, 10^{-2} M) the double layer has decreased to 9.62 nm. Third, add too much salt, and the double layer collapses. It is so small that attractive forces will, sooner or later, cause aggregation since, commonly, there are particle surface patches without charge, and these are ripe for aggregation sites.

Fourth, much smaller concentrations of multivalent ions can cause aggregation. Therefore, if working in tap water, it is important to know what the concentrations of the various ions are.

Most measurements are made at or near room temperature, which, for convenience, we define as 25 °C. It is then fair to ask what the variation in the calculated double layer thickness will be with temperature. The following, simple equation may be used to calculate the dielectric constant of water from t = 10 to t = 60 °C to within 0.25% of the actual value is:

$$\varepsilon_r = 78.54 \times 10^{-0.0022 \cdot (t-25)} \tag{Z2-3}$$

As temperature increases the dielectric constant decreases. These two effects tend to offset each other in the calculation of κ^{-1}. The slight decrease in density with increasing temperature causes a slight decrease in concentration and ionic strength, which further offsets the decrease in dielectric constant. Finally, κ^{-1} varies as the square root of all these changes. Putting numbers in shows that κ^{-1} varies by less than +1% from 10 to 40 °C, respectively. Therefore, the values calculated at 25 °C will be enough for most purposes regardless of the actual measurement temperature.

NOTE: Does ζ potential even have a meaning when the double layer thickness is smaller than a few Angstroms? Such is the case when the salt concentration is $\geq$ 100 mM (10^{-1} M). At that point, electrophoretic mobility still has a definition: electrophoretic velocity divided by applied electric field strength. And it can be used to describe charged particles in solution or suspension. But ζ potential is probably not a useful concept at that point. See end of Appendix Z1.

Appendix Z3: The Smoluchowski Equation and the Debye - Hückel Approximation

Smoluchowski Equation

In the Smoluchowski limit the electrophoretic mobility μ_e is related to the zeta potential by the simple equation:

$$\mu_e = \varepsilon \cdot \zeta / \eta \qquad \text{(Z3-1)}$$

Where,

$$\varepsilon \ = \ \text{Permittivity of Liquid}$$
$$\zeta \ = \ \text{Particle Zeta Potential}$$
$$\eta \ = \ \text{Viscosity of Liquid}$$

In Appendix Z2 it was stated that ε for H_2O at 25 °C is the product of ε_r, the dielectric constant (78.54 for water), and the permittivity of vacuum ε_0, 8.8542 x 10^{-12} J/($V^2 \cdot$m). The viscosity of water at 25 °C is equal to 0.8904 x 10^{-2} g/(cm·s).

One mobility unit is 1 $\mu \cdot s^{-1} \cdot V^{-1} \cdot cm = 10^{-8}$ $m^2 \cdot s^{-1} \cdot V^{-1}$. Therefore, the zeta potential in water at 25 °C corresponding to one mobility unit is equal to ζ = 12.8 mV per mobility unit. (See if you can calculate that using the right units and conversions.)

Debye-Hückel Approximation

One of the assumptions in Henry's equations is that the potential is small. This same assumption occurs in several other theories of the diffuse double layer, and it has roots in the Debye-Hückel approximation:

$$\Psi_o = k_B \cdot T / e << 1 \qquad \text{(Z3-2)}$$

Where,

$$k_B \ = \ \text{Boltzmann's Constant, } 1.3807 \text{ x } 10^{-23} \text{ J·K}^{-1}$$
$$T \ = \ \text{Temperature in Kelvin, K}$$
$$e \ = \ \text{Electronic Charge in Coulombs, } 1.6022 \text{ x } 10^{-19} \text{ C}$$

When evaluated at 25 °C, Ψ_o = 25.7 mV. (See if you can calculate that, remembering if you move a coulomb of charge across one volt you need one joule of energy.) In other words, Henry's equation is increasingly more accurate as the zeta potential falls below 25.7 mV. And this corresponds to about 2 mobility units.

Appendix Z4: Errors in Calculating Zeta Potential

The Smoluchowski equation for calculating zeta potential from the measured mobility is so common place that one forgets that the equation is only an approximation. It is a good approximation when $\kappa \cdot a \gg 1$, where "a" is the particle radius, often approximated by the hydrodynamic radius r_H. Often, however, $\kappa \cdot a$ is between 10 and 100. It is worthwhile estimating the errors between the true zeta potential and that calculated using the Smoluchowski and Henry equations.

The tables below take advantage of interpolating values shown in figures from the original work of Wiersma, Loeb and Overbeek.[51] For the true zeta potential Wiersma calculated the correction factor $f(\kappa \cdot a, \zeta)$ including relaxation and retardation effects. The first column is the true zeta potential. The second column shows the electrolyte. The third column was interpolated from referenced figures at $\kappa \cdot a = 10$. The fourth column is the calculated zeta potential using the Smoluchowski equation. The fifth column shows the percent error using the Smoluchowski equation. The second table uses Henry's equation for a 1:1 electrolyte to estimate the zeta potential. (The 2:1 electrolyte is not applicable since Henry only gave results for the 1:1 case.)

For $\kappa \cdot a = 10$ in H_2O at 25°C:

Errors in Calculating Zeta Potential: Smoluchowski

ζ_T(mV)	Elect.	$f(\kappa \cdot a, \zeta)$	ζ_{Smol}(mV)	%Err
25.7	1:1	1.238	31.8	+24
51.4	1:1	1.143	58.7	+14
51.4	2:1	0.833	42.8	- 17

Errors in Calculating Zeta Potential: Henry

ζ_T(mV)	Elect.	$f(\kappa \cdot a, \zeta)$	ζ_{Henry}(mV)	%Err
25.7	1:1	1.238	28.3	+10
51.4	1:1	1.143	56.7	+10
51.4	2:1	0.833	N.A.	N.A.

In summary, the Smoluchowski equation can lead to errors as large as +24% for $\kappa \cdot a = 10$. Henry's equation is the better choice, with errors as large as +10% for $\kappa a = 10$. One needs, however, an estimate of the average particle radius to apply the Henry equation; therefore, it only applies to monodisperse size distributions.

---

[51] Wiersema, P. H., Loeb, A. L. and Overbeek, J. Th. G., Journal of Colloid and Interface Science, vol. 22, p. 78, (1966).

Appendix Z5: Electrophoretic Doppler Shift

The shift in frequency, $\Delta\omega_s$, of scattered light from a charged particle moving with electrophoretic velocity, $\mathbf{V_e}$, in an electric field, $\mathbf{E}$, is given by the dot product of vectors (bold for vectors, not bold for scalars):

$$\omega_s = \mathbf{q}\cdot\mathbf{V_e}, = q\cdot V_e\cdot\cos\phi \qquad\qquad (Z5\text{-}1)$$

Where, ϕ is the angle between the vectors. When the electrical field is perpendicular to the incident laser beam direction, represented by the unit vector $\mathbf{k_o}$, then it can be shown that ϕ is related to the scattering angle θ by $\phi = \theta/2$ for the special case when $\mathbf{E} \perp \mathbf{k_o}$.

The magnitude of $\mathbf{q}$, the scattering wave vector, is given by $q = (4\pi n/\lambda_0)\sin(\theta/2)$ where n is the refractive index of the liquid and λ_0 is the wavelength of the laser in vacuum. For water, n = 1.332 in the red end of the visible range of wavelengths and doesn't change much with λ_0.

Since $\mathbf{V_s} = \mu_e\cdot\mathbf{E}$ where μ_e is the electrophoretic mobility, the equations and parameters above, using λ_0 = 640 nm = 0.670 x 10^{-4} cm and $\theta = 15°$ (typical for modern ELS instruments to suppress diffusion broadening, see Appendix Z6), combine to give:

$$\omega_s = 3.38\cdot\mu_e\cdot E \text{ in rad/s,} \qquad \text{or} \qquad \nu_s = \omega_s/2\pi = 0.539\cdot\mu_e\cdot E \text{ in Hz} \qquad (Z5\text{-}2a, \text{-}2b)$$

Where, E is the magnitude of the electric field vector in units of V/cm and μ_e is in $(\mu/s)/(V/cm)$.

When the mobility is low, high field strength is needed to make good measurements, and when the mobility is high, low field strength is needed.

Indirectly, conductivity also plays a part: At high conductivity, it is difficult to establish high field strength. Nevertheless, a reasonable approximation for the maximum of the product $\mu_e\cdot E$ is 400 μ/s. Then a maximum Doppler shift is $\nu_s \sim$ 200 Hz. In fact, at low scattering angles like $\theta = 15°$, ν_s varies from 0 to ±100 Hz. These Doppler shifts due to the electrophoretic mobility are added to or subtracted from the applied frequency shift (to determine sign of V_e, μ_e and ζ). The reference beam shift in a Brookhaven Instruments NanoBrook ZetaPlus or ZetaPALS is 250 Hz. If the measured frequency shift is less than 250 Hz, then the calculated electrophoretic mobility and resultant zeta potential must be negative; whereas, if the measured frequency shift is greater than 250 Hz, then V_e, μ_e and the ζ potential must be positive.

Appendix Z6:   Diffusion & Thermal Broadening Effects

Diffusion Broadening

The velocity distribution of charged particles moving between two electrodes is broadened not only by the electrophoretic mobility distribution of interest but also by the diffusional broadening due to particle size. This latter broadening is, in fact, the principle used for measuring particle size with DLS. The frequency of the scattered laser line is broadened by the diffusional motion of the scattering particles. The half-width at half-height of the power spectral density of the scattered electric field determined using a spectrum analyzer, is called the linewidth. The linewidth designated Γ is given by:

$$\Gamma = D_T \cdot q^2 \text{ in rad/s} \tag{Z6-1}$$

Where, D_T is the translational diffusion coefficient and q is the magnitude of the scattering wave vector (See Appendix Z5 for the definition of q). The diffusion coefficient is related to particle size by:

$$D_T = k_B \cdot T/(3 \cdot \pi \cdot \eta \cdot d_H) \text{ in cm}^2/\text{s} \tag{Z6-2}$$

This is the famous Stokes-Einstein equation, where

$$k_B \quad = \quad \text{Boltzmann's Constant, } 1.3807 \times 10^{-23} \text{ J} \cdot \text{K}^{-1}$$

$$T \quad = \quad \text{Temperature in Kelvin, K}$$

$$\eta \quad = \quad \text{Viscosity of Liquid, g/cm} \cdot \text{s}$$

$$d_H \quad = \quad \text{Hydrodynamic diameter, cm}$$

Hydrodynamic diameter requires further explanation. It includes the geometric size of the particles plus whatever is attached to the surface such as solvents molecules, surfactants, wetting agents, and dispersing agents. d_H is the equivalent hydrodynamic diameter of the particle with the same diffusion coefficient as the one measured. In many situations, it is close to the geometric size, but this is not always true. A classic case where it is not is a small diameter particle stabilized by a large molecular weight, and therefore extended, polymer.

For measurements made at 25 °C (298.15 K) in water, $\eta = 0.8904$ cP, where one Poise has units of $g \cdot cm^{-1} \cdot s^{-1}$. Substituting the parameters in Eq. (Z6-2) into Eq. (Z6-1) plus the equation for q given in Appendix Z5, yields:

$$\Gamma_{15°}(\text{rad/s}) = 5716/d(\text{nm}) \quad \text{and} \quad \Gamma_{90°}(\text{rad/s}) = 167,800/d(\text{nm}), \quad \text{or}$$

$$\Gamma_{15°}(\text{Hz}) = 909.7/d(\text{nm}) \quad \text{and} \quad \Gamma_{90°}(\text{Hz}) = 26,710/d(\text{nm}). \tag{Z6-3a, -3b}$$

The following table gives some typical examples:

d(nm)	$\Gamma_{15°}$ (Hz)	$\Gamma_{90°}$ (Hz)
10	91.0	2,670
20	45.5	1,340
50	18.2	534
100	9.1	267
200	4.5	134
500	1.8	53.4
1,000	0.9	26.7
2,000	0.5	13.4

Table Z6-1: Linewidths due to diffusional broadening as a function of size and angle.

There are two lessons to be learned from these numbers. First, they show the advantage of making electrophoretic mobility measurements at low angles. The broadening due to translational diffusion is much less at low angles than at high angles. Therefore, the shift and the broadening due to the mobility distribution are easier to extract from the total broadening at low angles. Second, for small particles it becomes increasingly difficult to separate the measured broadening into mobility and particle size effects. In Appendix Z5 it was stated that typical Doppler shifts will be less than ~100 Hz. Depending on the actual shift, the diffusion broadened component for small particles can exceed the mobility shift. Under these circumstances the uncertainty in the peak of the measured frequency distribution becomes greater, and the measured width is dominated by the diffusion effect.

NOTE: The above analysis applies to ELS, electrophoretic light scattering. It does not apply to PALS. To first order, diffusional broadening plays no part in PALS. Thus, for the smallest particles PALS is a better choice than ELS. Monodisperse, globular proteins should on this account be measured using PALS. Since they are often in saline, a relatively high salt condition, PALS is also preferred due to its sensitivity and the lack of need for high field strength in conductive media.

Thermal Broadening

Several other effects also broaden the measured distribution. Of these, thermal effects are the most significant. Electrophoretic velocities are on the order of tens of microns per second. But so are thermal drifts. And they can be in random directions, thus broadening the measured distribution. Or they can add to or subtract from the electrophoretic velocity if they pile up in one direction.

To reduce thermal effects, use an instrument with temperature control. And wait after sample insertion for equilibrium. When using ELS, this amounts to waiting for the central peak, with field off, to remain within a few hertz of the reference beam frequency.

Electro-osmotic effects add a systematic shift to measurements if they are present. For this reason, classic cells required measurement at the stationary plane. This effect is eliminated in a modern cell such as that found in the Brookhaven instruments. However, there is one exception if surface zeta potential is being measured. See Chapter VI. From such surface (macroscopic surface) zeta potential determinations, if the surface is more than about one millimeter from the beam, then such affects are insignificant. This puts a limit on how small zeta potential cells can be made.

Appendix Z7: Force Between Two Charged Particles

Start with Coulomb's Law for the force between two charges q_1 and q_2:

$$F = \frac{q_1 \cdot q_2}{4 \cdot \pi \cdot \varepsilon \cdot r^2} \qquad \text{Z7-1)}$$

Where $\varepsilon = \varepsilon_o \cdot \varepsilon_r$ with ε_o, the permittivity of free space, equal to 8.8.8542 x 10^{-12} Farad/meter, and ε_r the relative permittivity or dielectric constant for the medium between the two charges, and r the distance between the charges.

The electrostatic potential ψ due to a point charge is:

$$\psi = \frac{q}{4 \cdot \pi \cdot \varepsilon \cdot r} \qquad \text{(Z7-2)}$$

The potential of a sphere of radius r with total surface charge $Q_s = e \cdot Z$, where e = 1.6022 x 10^{-19} Coulombs and Z (positive or negative integer) the total number of charges is similarly given by:

$$\psi_s = \frac{Q_s}{4 \cdot \pi \cdot \varepsilon \cdot r} \qquad \text{(Z7-3)}$$

And at the shear plane, where r is the hydrodynamic radius r_H, and where the zeta potential is defined, a similar equation yields the total charge Q_{sp} at the shear plane:

$$\zeta = \frac{Q_{sp}}{4 \cdot \pi \cdot \varepsilon \cdot r_H} \qquad \text{(Z7-4)}$$

The force of repulsion is proportional to ψ^2 for point charges. And, as charged spheres approach each other, the force of repulsion is proportional to ζ^2. When electrophoretic mobility μ_e is strictly proportional to the ζ potential, as it is in the Smoluchowski and Hückel limits only, then the force of repulsion is also proportional to μ_e^2. In all other cases it is not. For this reason, using μ_e as a measure of stability is only true in these limiting cases.

Stability of dispersions is a balance between attractive and repulsive forces. Attractive forces, such as van der Waal's, are always present. Repulsive forces can be changed, most notably by changing the surface charge density on the particle. Changing the pH, adding surfactants, wetting and dispersing agents may change the surface charge density. These changes are monitored by a determination of the zeta potential. It is for this reason that zeta potential is such a useful technique.

Appendix S1: Particle Counting and Statistics[52]

To calculate the number of particles, n, needed to establish a certain percent standard error (standard deviation of the mean) σ for a narrow distribution, use the result from Poisson statistics, namely:

$$\sigma = 100/\sqrt{n}$$

σ	n
1%	10,000
2%	2,500
5%	400

To get a 1% standard error on the mean value, you need to count 10,000 particles. Accepting a modest 2% error requires counting 2,500 particles. If you are only interested in a 5% error, you only need to count 400 particles. All this is statistical, assuming random error only. So, if you systematically miss some size classes (perhaps focusing problems in image analysis or large, dense particles falling out of a stream), that is another type of error, perhaps more serious than the statistical counting error.

In any case, you need to count a lot of particles. They can't be touching. Thus, you need to dilute the sample. They can't be partially in and partially out of a field of view if you are doing image analysis, because you then don't know how big or small, they really are. All particles touching the edges of a field must be ignored. The result is that you need to measure a lot of fields and a lot of particles to do number counting properly.

As the distribution gets broader, even more particles need to be counted. Assuming the distribution is Gaussian, and not too many are, one can use the definition of the relative standard error SE as it relates to the relative standard deviation s (a relative measure of the distribution width):

$$SE = s/\sqrt{n}$$

Squaring and rearranging we obtain the result for distributions that look Gaussian and for which we have an estimate of the width s (perhaps from initial measurements):

$$n = s^2/SE^2$$

Suppose a preliminary measurement produced a relative standard deviation (standard deviation divided by the mean) of 3, a relatively broad distribution. And you were aiming for a 2% error on the mean, so SE = 0.02. How many particles should you count?

---

[52] See Particle Size Measurement, T. Allen, 3rd Edition, Chapman & Hall, 1981 (or one of the later editions) pages 210-211.

The answer is: n= $3^2/(0.02)^2$ = 22,500. It is easy to see how quickly the number increases the broader the distribution.

More advanced approaches are described and referenced in the ISO standard 13322-1.[53]

But the end results tell the same story: Except for the narrowest of distributions, you must count at least several hundred particles to get a small error bar on the mean, and tens of thousands of particles for broad distributions. This, again, assumes no experimental bias against large, dense, or small particles.

---

[53] ISO 13322-1, Particle Size Analysis—Image Analysis Methods Part 1: Static Image Methods. Available at www.iso.org.

Appendix S2: Exact Solutions for Stokes' Law

Sedimentation

The solutions to the equations of motion for sedimentation and centrifugation were obtained by assuming acceleration went to zero quickly and the velocity reached the terminal velocity rapidly. Under these conditions the linear, 2nd order, homogeneous differential equations with constant coefficients were reduced to 1st order differential equations which could be easily integrated. The purpose here is to provide the general solutions and see how quickly they produce terminal velocities and thus validate the assumptions.

Starting with sedimentation, Newton's equation of motion is:

$$m_p \ddot{x} = m_p g - m_L g - f_D \dot{x} \tag{S2-1}$$

Limiting conditions are the following: x = 0 when t = 0; $\dot{x} = 0$ when t = 0.

The general solution is: $x = \dfrac{g \cdot m_p \cdot \Delta m}{f_D^2} \left[\dfrac{f_D \cdot t}{m_p} + e^{-\frac{f_D \cdot t}{m_p}} - 1 \right] \to 0$ when t $\to$ 0 $\tag{S2-2}$

Taking the derivative yields this: $\dot{x} = \dfrac{g \cdot \Delta m}{f_D} \left[1 - e^{-\frac{f_D \cdot t}{m_p}} \right] \to 0$ when t $\to$ 0 $\tag{S2-3}$

This last equation yields the terminal velocity $\dot{x}_T = \dfrac{g \cdot \Delta m}{f_D}$ when t $\to \infty$. And when integrated, the terminal velocity leads to Stokes' Law of Sedimentation under gravity. How fast this terminal velocity is reached will be detailed below. But for now, equation S2-3 can be used to obtain the acceleration:

$$\ddot{x} = \dfrac{g \cdot \Delta m}{m_p} \cdot e^{-\frac{f_D \cdot t}{m_p}} \tag{S2-4}$$

Plugging equations S2-4 and S2-3 into equation S2-1 yields an identity, demonstrating that equation S2-2 is the general solution.

So how fast does $exp\left(-\dfrac{f_D t}{m_p}\right)$ go to zero? To find out, substitute for $f_D = 3\pi \eta_L d_p$ and, $m_p = \rho_p (\frac{\pi}{6}) d_p^3$. The result is: $\dfrac{f_D}{m_p} = \dfrac{18 \cdot \eta_L}{\rho_p \cdot d_p^2}$. The exponential dies away more slowly the smaller this ratio. So, pick, for example a large particle size, say 100 μ (0.01 cm), a large density, say 10 g/cm³, and a low viscosity, say 1 mPa·s. In a real experiment, given how large and dense the particle is, one would opt for a much higher viscosity to slow it down, but here we are looking for limiting cases.

Substituting gives $(18 \cdot 10^{-2}\,\text{g/cm·s})/(10\,\text{g/cm}^3 \cdot (0.01\,\text{cm})^2) = 180\,\text{s}^{-1}$. Table S2-1 shows values for the difference between the true velocity and the terminal velocity after fixed times. Column 4 represents the percent difference.

Time	$\exp(-f_D t/m_p)$	$1 - \exp(-f_D t/m_p)$	% Difference from $\dot{x}_T$
1 ms	0.835	0.165	83.5%
10 ms	0.165	0.835	16.5%
100 ms	1.5×10^{-8}	1	0
1 s	6.7×10^{-79}	1	0

Table S2-1: How fast does velocity approach terminal velocity with gravity?

By 100 ms the terminal velocity is reached. And this was for an extreme case. For all other cases, it is reached between 100 ms and 10 ms. Thus, setting $\ddot{x}_T$ to zero to calculate the final Stokes' Law was more than justified. It also tells us that even if the particle had an initial non-zero velocity it would quickly reach terminal velocity, at least in a liquid.

Centrifugation

Turning (pun intended) to centrifugation, Newton's equation of motion is:

$$m_p \ddot{r} = m_p \omega^2 r - m_L \omega^2 r - f_D \dot{r} \qquad \text{(S2-5)}$$

Boundary conditions are the following: $r = r_s$ when $t = 0$; $\dot{r} = 0$ when $t = 0$; from which it follows from equation S2-5 that $\ddot{r} = \dfrac{\omega^2 \Delta m \cdot r_s}{m_p}$ when $t = 0$.

Start with a trial solution: $\qquad r = Ce^{at}; \dot{r} = aCe^{at}; \ddot{r} = a^2 Ce^{at} \qquad$ (S2-6a, -6b, -6c)

Plug these into equation S2-5 and divide by Ce^{at} to yield:

$$m_p a^2 = \omega^2 \Delta m - f_D a,$$

Which can be written as: $\qquad m_p a^2 + f_D a - \omega^2 \Delta m = 0. \qquad$ (S2-7)

This is a quadratic equation with two solutions:

$$a_1 = -\frac{f_D}{2 \cdot m_p} + \frac{f_D}{2 \cdot m_p}\sqrt{\left(1 + \frac{4\omega^2 m_p \Delta m}{f_D^2}\right)} \quad \&$$

$$a_2 = -\frac{f_D}{2 \cdot m_p} - \frac{f_D}{2 \cdot m_p}\sqrt{\left(1 + \frac{4\omega^2 m_p \Delta m}{f_D^2}\right)} \qquad \text{(S2-8a, -8b)}$$

The general solution is then given by:

$$r = C_1 e^{a_1 t} + C_2 e^{a_2 t},$$

and from the 1st boundary condition it follows:

$$r_s = C_1 + C_2 \qquad (S2-9)$$

$$\dot{r} = a_1 C_1 e^{a_1 t} + a_2 C_2 e^{a_2 t},$$

and from the 2nd boundary condition it follows

$$0 = a_1 C_1 + a_2 C_2 \qquad (S2-10)$$

The boundary conditions resulting in equations S2-9 and S2-10 at t = 0 can be used to express C_1 and C_2 as:

$$C_1 = -\frac{a_2 r_s}{a_1 - a_2} \quad and \quad C_2 = +\frac{a_1 r_s}{a_1 - a_2} \qquad (S2-11a, -11b)$$

Substituting equations S2-11a and S2-11b into equation S2-10 gives:

$$r = \frac{-a_2 r_s}{a_1 - a_2} e^{a_1 t} + \frac{a_1 r_s}{a_1 - a_2} e^{a_2 t} \quad and \quad \dot{r} = \frac{-a_1 a_2 r_s}{a_1 - a_2} e^{a_1 t} + \frac{a_1 a_2 r_s}{a_1 - a_2} e^{a_2 t} \qquad (S2-12a, -12b)$$

To estimate the terminal velocity, we must investigate what happens when t → ∞. Since both terms of a_2 are negative, $e^{a_2 t}$ goes to zero. Since the first term of a_1 is negative, only its second term, $\frac{f_D}{2 \cdot m_p} \sqrt{\left(1 + \frac{4\omega^2 m_p \Delta m}{f_D^2}\right)}$, contributes. Thus, for long times, the radial position and terminal velocity are reduced to

$$r = \frac{-a_2 r_s}{a_1 - a_2} e^{\frac{f_D \cdot t}{2 \cdot m_p} \sqrt{\left(1 + \frac{4\omega^2 m_p \Delta m}{f_D^2}\right)}} \quad and \quad \dot{r}_T = \frac{-a_1 a_2 r_s}{a_1 - a_2} e^{\frac{f_D \cdot t}{2 \cdot m_p} \sqrt{\left(1 + \frac{4\omega^2 m_p \Delta m}{f_D^2}\right)}} \qquad (S2-13a, -13b)$$

Substituting equation S2-13a into equation S2-13b takes us a step closer to Stokes' Law as derived from the terminal velocity. The velocity after a long time is given by:

$$\dot{r}_T = a_1 \cdot r \qquad (S2-14)$$

This has the correct form such that when integrated t = $1/a_1 \cdot$ Ln(R_D/r_s), where R_D is the radial position of the detector. It remains to be seen if a_1 as given by equation S2-8a can be simplified.

Is $\dfrac{4\omega^2 m_p \Delta m}{f_D^2}$ about equal to 1, much less than 1, or much greater than 1? We need typical

values. First, substitute for m_p, Δm, and f_D: $m_p = \rho_p\left(\frac{\pi}{6}\right)d_p^3$, $\Delta m = \Delta\rho \cdot \left(\frac{\pi}{6}\right)d_p^3$, and

$f_D = 3\pi\eta_L d_p$.

The result is this:

$$\frac{4\omega^2 m_p \Delta m}{f_D^2} = \frac{d_p^4 \cdot \omega^2 \cdot \Delta\rho \cdot \rho_p}{81 \cdot \eta_L^2} \tag{S2-15}$$

Evaluating this expression for a variety of circumstances is shown in Table S2-2.

Particle/liquid	ω (RPM)	d_p (nm)	ρ_p (g/cm³)	ρ_L (g/cm³)	η_L (mPa·s)	$\dfrac{d_p^4 \cdot \omega^2 \cdot \Delta\rho \cdot \rho_p}{81 \cdot \eta_L^2}$
PS/water	1,500	5,000	1.051	0.997	0.933	1.24×10^{-8}
TiO$_2$/water	1,500	500	3.910	0.997	0.933	2.49×10^{-10}
Cb Blk/water	8,000	100	1.84	0.997	0.933	1.54×10^{-12}
Fe$_2$O$_3$/water	12,000	50	5.24	0.997	0.933	3.11×10^{-12}

Table S2-2: Simplifying a_1.

The values in the last column show that the 2nd term under the square root sign (equations S2-13a, and -13b) is very small. This allows a Taylor's expansion resulting in:

$$a_1 = -\frac{f_D}{2 \cdot m_p} + \frac{f_D}{2 \cdot m_p}\sqrt{\left(1 + \frac{4\omega^2 m_p \Delta m}{f_D^2}\right)} \rightarrow \omega^2 \Delta m / f_D \tag{S2-16}$$

Substituting into equation S2-14 gives $\dot{r}_T = \dfrac{\omega^2 \Delta m}{f_D} \cdot r$. When integrated this leads to Stokes'

Law in a centrifuge. The question remaining is how fast does $\dot{r} \rightarrow \dot{r}_T$? The argument leading to equations S2-13a and S2-13b was that $e^{a_2 t}$ went to zero very fast. Using the same arguments used in equation S2-16, it follows that:

$$a_2 = -\frac{f_D}{2 \cdot m_p} - \frac{f_D}{2 \cdot m_p}\sqrt{\left(1 + \frac{4\omega^2 m_p \Delta m}{f_D^2}\right)} \rightarrow -\frac{f_D}{m_p} - \omega^2 \Delta m / f_D \tag{S2-17}$$

Substituting we have $\dfrac{f_D}{m_p} = \dfrac{18 \cdot \eta_L}{\rho_p \cdot d_p^2}$ with values shown in the last column in Table S2-3.

Sample/ Water	ω (RPM)	d_p (nm)	ϱ_p (g/cm^3)	η_L (mPa·s)	ρ_L (g/cm^3)	$\dfrac{\omega^2 \Delta m}{f_D}$	$\dfrac{f_D}{m_p}$
PS	1,500	5,000	1.051	0.933	0.997	1.98 x 10^{-3}	6.40 x 10^{+5}
TiO2	1,500	500	3.910	0.933	0.997	1.07 x 10^{-3}	1.72 x 10^{+7}
Carbon Black	8,000	100	1.84	0.933	0.997	3.52 x 10^{-4}	9.12 x 10^{+8}
Fe$_2$O$_3$	12,000	50	5.24	0.933	0.997	9.97 x 10^{-4}	1.28 x 10^{+9}

Table S2-3: Evaluating f_D/m_P under a variety of samples and run conditions.

And substituting we have $\dfrac{\omega^2 \Delta m}{f_D} = \dfrac{\omega^2 \Delta \rho d_p^2}{18 \cdot \eta_f}$ with values shown in second to last column of

Table S2-3. In every case the term with $-\dfrac{f_D}{m_p}$ is completely dominant. For $\dot{r} \to \dot{r}_T$, the sec-

ond term in equation S2-12b must go to zero. That means $e^{-\frac{f_D \cdot t}{m_p}} \to 0$, just like in equation S2-4 in the case of gravity, except here the particles are much smaller and the terminal velocity reached much sooner.

Sample/Liquid	$\dfrac{f_D}{m_p}$	$exp(-f_D \cdot t / m_p)$		
		1 μs	10 μs	100 μs
PS/water	6.40 x 10^{+5}	0.53	1.66 x 10^{-3}	0
TiO2/water	1.72 x 10^{+7}	0	0	0
Carbon Black/water	9.12 x 10^{+8}	0	0	0
Fe$_2$O$_3$/water	1.28 x 10^{+9}	0	0	0

Table S2-4: How fast does velocity approach terminal velocity with centrifugation?

The results show that in less than 10 μs all results have reached terminal velocity. Again, this validates what an excellent approximation it is to assume the acceleration is zero and solve the first order differential equation. And it again shows how fast a particle reaches mechanical equilibrium after injection. Even if the injected particles smash into the back wall of the spinning disc and are hurtled downward towards the meniscus, they reach terminal velocity so fast compared to the few hundred seconds of even a fast measurement.

Appendix S3: Upper Size Limit for Sedimentation and Centrifugation: Low Reynolds Number

Stokes' Law holds assuming laminar flow: orderly flow around particles as they move. This requires no turbulence. The Reynolds number, Re, describes regions of flow and the onset of turbulence. When Re ≤ 0.2, errors calculated assuming laminar flow are less than 2% for the maximum particle size, d_{max}. This is useful for calculating maximum sizes when using either gravitational sedimentation or centrifugation.

The Reynolds number is given by the following:
$$Re = \frac{d_p \cdot v \cdot \rho_L}{\eta_f}. \tag{S3-1}$$

Where, η_f and ρ_f are the liquid viscosity and density, and v is the particle velocity.

Case I. Gravitational sedimentation

The terminal velocity for this case is given by, $v = \dot{x}_T = \dfrac{g \cdot \Delta m_p}{f_D} = \dfrac{g \cdot \Delta \rho \cdot d_p^2}{18 \cdot \eta_L}.$ (S3-2)

Substituting equation S3-2 into equation S3-1 and the condition Re ≤ 0.2, one finds a limit

$d_p = d_{max}$ on size for using Stokes' Law: $\quad d_{max} \leq \left[\dfrac{3.6 \cdot n_L^2}{g \cdot \rho_L \cdot \Delta \rho} \right]^{\frac{1}{3}}$ (S3-3)

The difference between the particle and liquid density is $\Delta \rho = \rho_p - \rho_f$. The following table assumes a temperature of 20 °C for the spin fluid viscosity. The particles are representative of a class, not a specific material. But they cover a useful range of conditions. The gravity constant $g = 981$ cm/s².

Particle	ρ_p (g/cm³)	liquid	ρ_L (g/cm³)	η_L (mPa·s)	d_{max} (micron)
Latex beads	1.1	water	0.998	1.002 x 10⁻²	154
Glass beads	2.0	water	0.998	1.002 x 10⁻²	72
Clay	5.0	20% Glycol-Water	1.026	1.476 x 10⁻²	58
Metal	10.0	60% Glycol-Water	1.078	5.016 x 10⁻²	98

Table S3-1: Upper size limit, d_{max} in microns, for range of materials using sedimentation.

What general conclusions can be drawn? For denser particles you must resort to more viscous liquids to increase the maximum diameter. Of course, these are inconvenient to use (mixing, removing bubbles, and cleaning after the measurement). The upper limit around

150 μ is suitable for low density particles in low density liquids like water. Unless you are willing to resort to high viscosity liquids, the upper limit for higher density particles is somewhere in the 50 μ range.

Case II. Centrifugation

The terminal velocity for this case is given by

$$v = \dot{r}_T = \frac{\omega^2 \cdot r \cdot \Delta m_p}{f_D} = \frac{\omega^2 \cdot r \cdot \Delta \rho \cdot d_p^2}{18 \cdot \eta_L}. \tag{S3-4}$$

Substituting equation S3-4 into equation S3-1 and the condition Re ≤ 0.2, one finds a limit

$d_p = d_{max}$ on size for using Stokes' Law: $\quad d_{max} \leq \left[\frac{3.6 \cdot \eta_L^2}{\omega^2 \cdot r \cdot \rho_L \cdot \Delta \rho} \right]^{\frac{1}{3}} \tag{S3-5}$

Very similar to equation S3-3, but the fixed gravitational constant is replaced by the product of the square of the disc rotation speed and a radius inside the spin fluid. This latter value, r, depends on the spin fluid volume, the position of the detector, the disc dimensions and changes during a measurement. For a typical Brookhaven BI-DCP (Disc Centrifuge Photosedimentometer) setup, r is about 4.5 cm and varies by a few tenths of a centimeter. Given the cube root relationship to d_{max}, using 4.5 cm to estimate a result is enough.

The following table assumes a temperature of 20 °C for the fluid viscosity. The particles are representative of a class, not a specific material. But they cover a useful range of conditions.

Particle	ρ_p (g/cm³)	ω^* (rad/s)	liquid	ρ_L (g/cm³)	η_L (mPa·s)	d_{max} (μ)
Latex beads	1.1	1,571	water	0.998	1.002 x 10⁻²	6.8
Glass beads	2.0	837.8	water	0.998	1.002 x 10⁻²	13.2
Clay	5.0	418.9	20% Glycol-Water	1.026	1.476 x 10⁻²	6.2
Metal	10.0	209.4	60% Glycol-Water	1.078	5.016 x 10⁻²	16.8

*Corresponds to 15,000, 8,000, 4,000 and 2,000 rpm.

Table S3-2:Upper size limit, d_{max} in microns, for range of materials using centrifugation.

What general conclusions can be drawn? All the results are larger than 5 μ, so using the centrifuge below that ensures laminar flow. A worst case would be running at very high speed with a high-density particle in a low viscosity liquid: For example, 10 nm Au particles run at 15,000 rpm in water at 25 °C. Then d_{max} should be less than or equal to 0.42 μ or 420 nm. Using typical run conditions for such a small particle, after 30 minutes of run time the maximum size is 65 nm and the lowest 4 nm. Neither of these values is larger than 420 nm so no turbulence is generated.

A parting thought. With a very broad size distribution, if the largest particle creates turbulence, it can "entrap" small particles and distort the measurement with a shift towards a larger and narrower distribution than the true one. Just as there is a convenience limitation using centrifuges on total duration for broad distributions, the same could be said about turbulence. With broad distributions if the largest particles can cause turbulence, the measurement suffers.

Appendix S4: Lower Size Limit for Centrifugation and Mean Square Displacement

As particles get smaller and smaller, thermal motion (also called Brownian motion) begins to compete with distance travelled by centrifugal motion. We can use this idea to estimate a minimum size below which Stokes 'Law is not useful for calculating size.

In 1905, Einstein and others showed that there was a relationship between mean square displacement (MSD), indicated below as $\overline{\Delta r^2}$, and the translational diffusion coefficient D_T. The relationship is

$$\overline{\Delta r^2} = 2 \cdot D_T \cdot t, \tag{S4-1}$$

Here t is the duration of the experiment.

It was also shown by Einstein that the Stokes-Einstein equation for the diffusion coefficient of a sphere is,

$$D_T = \frac{k_B T}{3\pi \eta_L d_p} \tag{S4-2}$$

This is the exact same equation used in DLS for particle size determination. Here as there, k_B = Boltzmann's constant 1.38065×10^{-16} erg/K, T is absolute temperature in degree K, and, as before, η_L is the liquid viscosity (0.8905×10^{-2} mPa·s or g/cm·s for water at 25 °C).

To complete the calculation, one must assume: The root-mean-square displacement (RMSD) equals some small fraction f of the distance travelled from meniscus at r_s to the detector at r_D. So, applying Stokes' Law of Centrifugation to the measured time for the particle to register at the detector amounts to ignoring this small diffusional displacement.

The assumption allows one to write,

$$\sqrt{(\overline{\Delta r^2})} = f \cdot (r_D - r_s) \quad \text{or} \quad \overline{\Delta r^2} = f^2 \cdot (r_D - r_s)^2 \tag{S4-3}$$

Combining Eq. (S4-1), (S4-2) and (S4-3) yields,

$$f^2 \cdot (r_D - r_s)^2 = 2 \cdot \frac{k_B T}{3\pi \eta_L d_p} \cdot t \tag{S4-4}$$

Here t was initially described as the time that the particle was diffusing. But it is also the time it was centrifuging from meniscus to detector and so it follows Stokes' Law of Centrifugation. Substituting $t = \frac{18\eta_L Ln(^{r_D}/r_s)}{\omega^2 \Delta \rho d_p^2}$ and remembering that d_p is now the minimum size before diffusion has an effect, we can calculate d_{min} as follows:

$$d_{min} = \left(\frac{12 \cdot k_B \cdot T \cdot Ln(\frac{r_D}{r_s})}{\pi \cdot f^2 \cdot \omega^2 \cdot \Delta \rho \cdot (r_D - r_s)^2} \right)^{1/3} \tag{S4-5}$$

To get an idea of where this leads, assume a small fraction due to diffusion might be 2% or 5%. That is assume in one case f = 0.02 and in another f = 0.05. To estimate r_D and r_s, we need a typical disc's parameters. A common one is the Brookhaven BI-DCP with r_D = 4.84 cm, and r_s = 3.75, 4.33, or 4.05 cm depending on the volume of spin fluid. With large sizes and densities, to prevent too short a measurement, it is best to use a large volume and low rotational speed. With small sizes and modest densities, to prevent too long a measurement, it is best to use a small volume and high rotational speed. Examples of each are given in the table. The temperature and viscosity correspond to 25 °C and water. Column seven uses f = 0.05 and column eight 0.02.

*Corresponds to 1500, 4000, 8000 and 12000 rpm

Particle	$\Delta \rho$ (g/cm³)	ω^* (rad/s)	liquid	r_s (cm)	r_D (cm)	d_{min}(nm) f = 0.05	d_{min}(nm) f = 0.02
TiO$_2$	2.9	157	water	3.75	4.84	57	105
Carbon Blk	1.0	838	water	4.05	4.84	29	53
Au	18.1	419	60% glycol-water	4.33	4.84	20	37
Polystyrene	0.054	1,257	water	4.33	4.84	68	125

Table S4-1: Effect of diffusion on size in centrifugation measurement

What does this mean? It means that if you want to keep the error from diffusion small, just a few percent or less, do not believe in sizes much smaller than those given in columns seven and eight at the rotational speeds given. It means that running at higher rotational speed, assuming you still capture a good baseline in the beginning of the experiment, allows you to believe more in smaller sizes.

And remember that an error in the smaller sizes of a distribution has very little effect on the mean and other moments of the distribution. It mostly affects the few percent undersize values in the cumulative distribution.

ABOUT THE AUTHOR

I was fortunate to receive an undergraduate education at the University of California at Berkeley. It was an exciting place in the first half of the 1960's and had world-class professors, one of whom, a Nobel Prize winner in Physics, taught our statistical mechanics course even though we were in the chemistry department. Moving on to MIT, I was fortunate to work with Prof. Carl Garland in physical chemistry, though what we did was closer to chemical physics. There is not a lot of difference. Carl taught me to respect data, look at data until you wring every piece of information out of it, and not to accept slipshod results. MIT in the second half of the 1960's was one of the birthplaces of what became Dynamic Light Scattering, DLS. My lab was down the hall from students of Prof. George Benedek, one of the fathers of DLS. When it came time to graduate, he recommended me to Prof. Ben Chu at The State University of New York at Stony Brook for a postdoc position.

By 1970 when I joined his group, Ben was one of just a few researchers working in the field of Quasielastic Light Scattering, QELS, later mostly known as DLS. He had learned Static Light Scattering, SLS, during his time at Cornell as one of the last postdocs of Prof. Peter Debye.

For lecture requests contact: bbradleyw58@gmail.com

www.ingramcontent.com/pod-product-compliance
Lightning Source LLC
Chambersburg PA
CBHW050837220326
41598CB00006B/386